Robotics Applications Programming with ROBOTIS DREAM Systems

By Chi N. Thai

CNT Robotics LLC, Buford
2019

Author: Chi N. Thai

roboteer@comcast.net

www.cntrobotics.com

Robotics Applications Programming with ROBOTIS DREAM Systems (2019)

ISBN 978-0-9993918-2-2 (Paperback).

Library of Congress Control Number: 2019939670.

Published by CNT Robotics LLC, Buford, GA, U.S.A.

Printed by Amazon Kindle Direct Publishing.

To my parents for giving me life.

To Christine and Emily for their enduring love and support.

To Jenson and all my grand nieces and nephews.

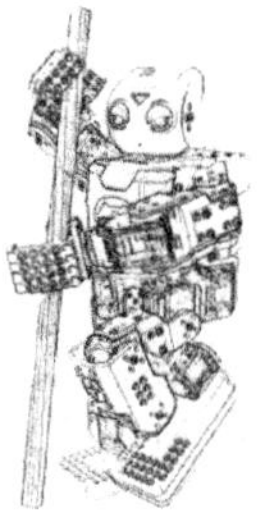

Preface

The DREAM II™ (School Set) programmable robotic kit was released by ROBOTIS® in Spring 2018. It is recommended for users at age 8 or older and is available from many on-line shops. It comes with instructions to build 23 programmable robot examples and it can be interfaced with two free popular programming tools: ROBOTIS TASK™ and MIT's SCRATCH® 2. The goal of my first DREAM II book "Learning Robotics with ROBOTIS DREAM Systems" is to provide a beginner in robotics with a firm foundation in robotics mechanical design, problem solving approaches and applications programming techniques using the above two software tools (https://www.amazon.com/Learning-Robotics-ROBOTIS-DREAM-Systems/dp/099939181X).

Who is this book for?

This second DREAM II book is designed to bring an advanced reader, already familiar with my first book, to higher levels of applications programming by using the "hidden" SMART DEVICE features of the ROBOTIS TASK tool allowing the robot's controller CM-150 to collaborate with the R+m.PLAY700 Mobile App (see video at https://www.youtube.com/watch?v=kAwDccoHC30), and also by applying the Edbot® Dream software tool to work with SCRATCH 2/3 and Python® (see video at https://www.youtube.com/watch?v=pOk5szsngM0. The Edbot Dream tool is particularly interesting as it allows multiple robots control from a single SCRATCH or Python program (see videos at https://www.youtube.com/watch?v=yUpx42eNR2I and https://www.youtube.com/watch?v=tQvGXYkiicw).

What do you need?

At the minimum, you need to purchase the ROBOTIS DREAM II School Set which is widely available via several web stores. Out of the box, this School Set is designed to be used within an MS Windows® compatible environment for robot program development with the free ROBOTIS TASK tool and for program download via a USB port using wired connection. It also includes a basic NIR-based remote control system (RC-100B).

The DREAM II School Set can also be used on mobile devices (phone or tablet), and the user would need to have the free Mobile Apps R+m.TASK and R+m.PLAY700 installed on these devices (available at (https://itunes.apple.com/us/developer/robotis-co-ltd/id948481761) or (https://play.google.com/store/apps/developer?id=ROBOTIS&hl=en). With this configuration, the user can create TASK programs that run on the Hardware Controller CM-150 and interact with various actuators and sensors, and that additionally access various sensing and multimedia services found on mobile devices. The user can configure the mobile device as a remote controller of the robot via a Bluetooth® receiver BT-210 or BT-410 which are additional purchases. The BT-210 works only with Windows and Android devices, while the BT-410 works only with iOS® and Android devices. To make the BT-410 work with Windows PCs, the user will need to additionally purchase the USB BT-410 Dongle.

If the user wants to use this DREAM II kit with the Edbot Dream software, it needs to be separately purchased from Robots in Schools Ltd (http://ed.bot/price-list) in the UK, or from other on-line shops. The Edbot Dream software runs on many platforms such as Windows PC, Mac and Raspberry Pi, and consequently it has many software interface options such as SCRATCH 2/3, Python and other computer languages are being developed. This book uses the Windows PC platform, the Off-Line SCRATCH 2 Editor (https://scratch.mit.edu/download) and the Thonny IDE (http://thonny.org/). Currently (Spring 2019) the Edbot Dream tool works with the ROBOTIS BT-210 receiver or the EDBOT Bluetooth Module (https://shop.ed.bot/collections/products/products/bluetooth-module-for-robotis-dream).

What is in this book?

This book consists of 2 independent chapters and uses selected DREAM II robots to showcase the capabilities of the software tools ROBOTIS TASK with SMART DEVICE commands, and EDBOT DREAM PRO:

1) Chapter 1 shows how to use the "hidden" SMART DEVICE features of the TASK tool to access various mobile device services such as how to display text and graphics, to use the touch areas, to perform text-to-speech and speech recognition tasks, to access multimedia devices such as musical instruments, video players and video cameras. Chapter 1 also shows how to use TASK and BlueTooth communications to achieve Master/Slave control of two CM-150 controllers.

2) Chapter 2 describes the software capabilities of the EDBOT DREAM PRO package and shows how to use SCRATCH 2/3 and Python with the DREAM II School Set. Applications using the PyGame, Threading, OpenCV-Python and NumPy modules are also showcased.

This book also provides appropriate source codes and tutorial videos (via YouTube®) to illustrate the presented concepts. Please visit http://www.cntrobotics.com/advanced-dream for access options to the source codes and tutorial videos.

And please, do let the author know about your experiences in using the materials in this book, this will help the author improve the next edition and the companion resources. Please visit the author web site www.cntrobotics.com for updates and other robotics materials. Please send your comments and suggestions to roboteer@comcast.net.

Trademark Attributions

Content in this book made references to the following trademarks:

ROBOTIS product names are trademarks or registered trademarks of ROBOTIS, Inc. Used with permission by ROBOTIS Co. Ltd.

Android®, Android Studio®, YouTube®, and Chrome® are registered trademarks of Google Inc.

SCRATCH® is a registered trademark of the Scratch Team at the Massachusetts Institute of Technology.

Windows®, Visual Studio Code® are registered trademarks of Microsoft Inc.

iOS®, iPhone® and iPad® are registered trademarks of Apple Corporation.

Bluetooth® is a registered trademark of Bluetooth SIG, Inc.

Edbot® is a registered trademark of Robots in Schools Ltd.

Python® is a registered trademark of the Python Software Foundation.

JavaScript® is a registered trademark of Oracle Corporation.

Buford, GA, USA Chi N. Thai

Contents

Chapter 1: Using R+m.PLAY700™ with TASK™

Currently (2019), ROBOTIS only advertises the use of the TASK tool with the DREAM II system (http://www.robotis.us/dream/), but Thai (2018) shows that users can also interface the DREAM II Controller (CM-150) to work with the "Windows PC / MIT SCRATCH 2" environment using the ROBOTIS R+SCRATCH helper application (V. 1.09 and above - http://www.robotis.us/roboplus2/). Furthermore, in this Chapter, readers will be shown how to interface a TASK program with the Mobile App called R+m.PLAY700 which was originally created to accompany only the PLAY700 kit (i.e. Controller CM-50) (http://www.robotis.us/software/play700/). But the PLAY700 App turns out to be applicable to all Firmware 2.0 Controllers, including the CM-150 (http://emanual.robotis.com/docs/en/dxl/protocol2/) and the interfacing is done by using SMART DEVICE commands in conjunction with SMART CONSTANT parameters inside the TASK tool. The "SmartDeviceControlTable.PDF" file contains detailed information about these SMART functions and their corresponding arguments, and this document is accessible at this web link (www.cntrobotics.com/advanced-dream). This PDF file is a Chrome-translated English version of this Korean web page
http://support.robotis.com/ko/software/mobile_app/r+smart/smanrt_manual.htm#Actuator_Address_0B3.

1.1 Installation and Usage of R+m.PLAY700

The latest version of R+m.PLAY700 can be downloaded and installed on your iOS or Android devices at these web links (https://play.google.com/store/apps/details?id=com.robotis.play700&hl=en or https://itunes.apple.com/us/app/play700/id1156037721?mt=8). The author had found that the Android's version of the PLAY700 App is more functional and less buggy than the iOS' version.

1.1.1 File & Folder Management Differences

For development work, the author happens to use a Windows 10 PC which interfaces well with Android devices, and all file and folder management can be done via Windows Explorer. Fig. 1.1 shows the folder/file structure for PLAY700 on an Android device.

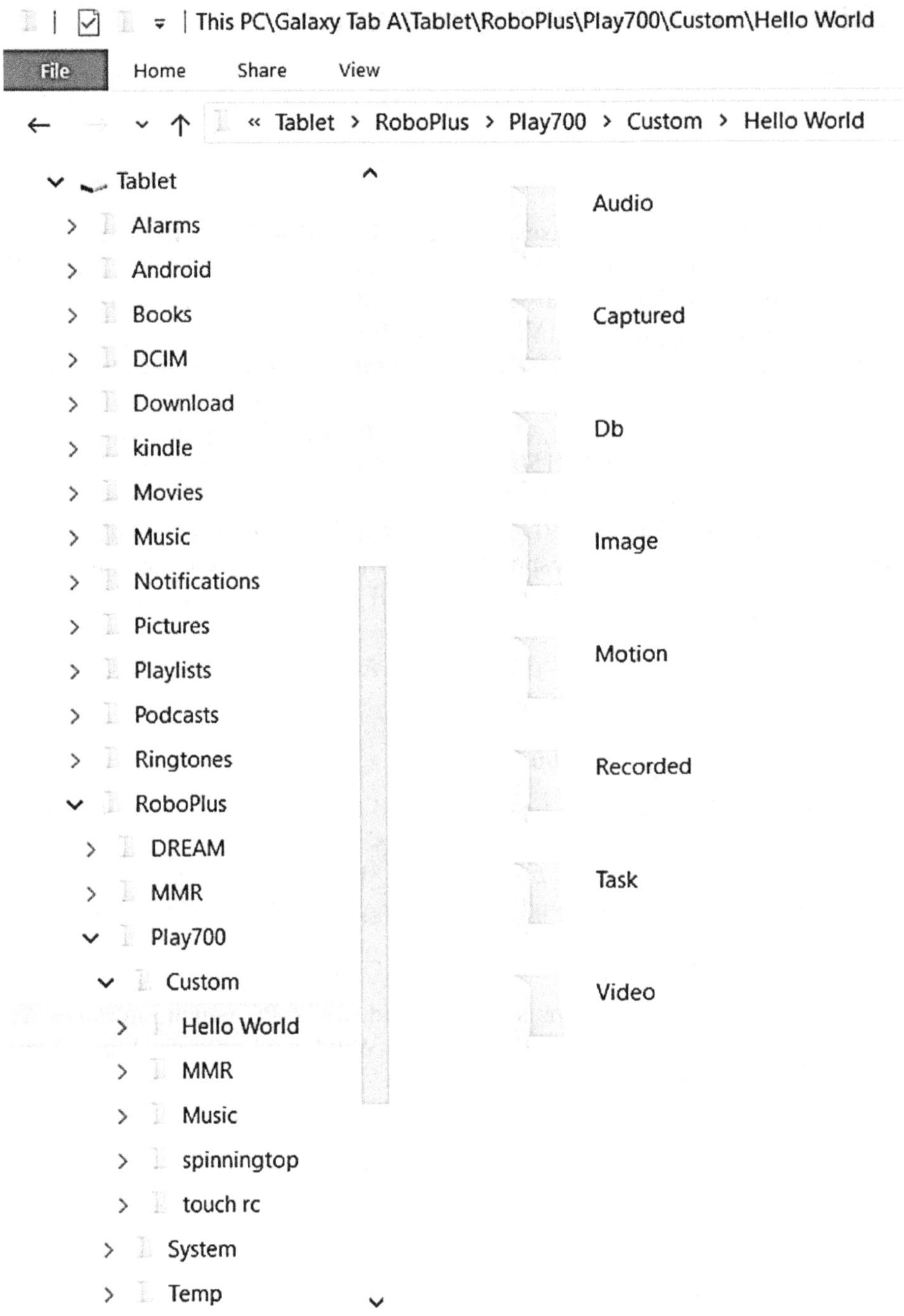

Fig. 1.1 File/Folder structure for PLAY700 App on an Android device.

For iOS devices, the author uses "iMazing™" to help with file/folder management for the PLAY700 App. Fig. 1.2 shows the folder/file structure for PLAY700 on an iOS device.

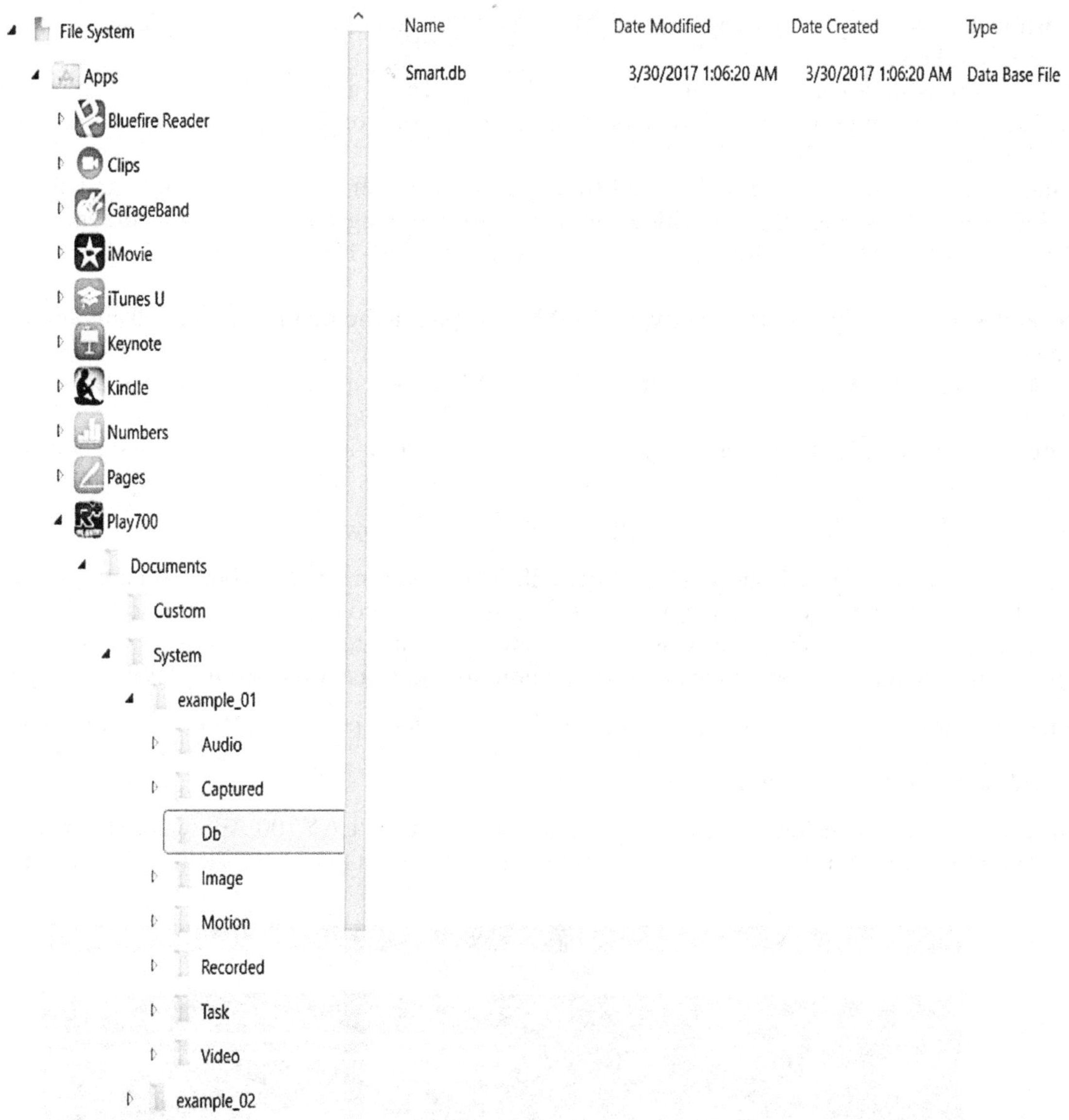

Fig. 1.2 File/Folder structure for PLAY700 App on an iOS device (via iMazing).

The reader can see that a PLAY700 project has many components represented by the folder structures shown in Figs. 1.1 and 1.2.

Fig. 1.1 shows that all ROBOTIS R+m apps are installed under a main folder named "RoboPlus", off the root directory of the main storage. As one drills into the PLAY700 folder, one can see three sub-folders "Custom", "System" and "Temp". The "System" sub-folder contains the 4 example projects provided by PLAY700. The "Custom" sub-folder contains the user-initiated projects, for example the "Hello World" project where the project's standard components are stored:

Audio – for audio files to be played by the mobile device but controlled by the TASK code at run time.

Captured – where pictures captured by the PLAY700 app via the mobile device's front and back cameras are kept.
Db – where SMART databases are stored, including "Text" and "Audio Input" lists used for Speech Recognition (these items can only be modified via the editing of the Project's Tools – see Sections 1.1.2 and 1.1.3).
Image – where photos specifically used by a given project are first "registered" and then "stored" as background images (i.e. "**Bg**" sub-folder) or as foreground images (i.e. "**Fg**" sub-folder).
Motion – where MTNX (Motion) files are stored. The CM-150 is not capable of processing Motion files.
Recorded - where videos captured by the PLAY700 app via the mobile device's front and back cameras are kept.
Task – where TSKX files are stored if the R+m.TASK tool is used to create TASK codes on the mobile device.
Video – for video files to be played by the mobile device but controlled by the TASK code at run time.

The "Temp" folder is only used by the PLAY700 App for its own purposes.

For an iOS device, Fig. 1.2 shows that ROBOTIS R+m Apps are installed under a main folder named "Apps" off the root directory of the main storage. As one drills into the PLAY700 folder, one would encounter a "Documents" sub-folder which further branches out into two sub-folders "Custom" and "System" which contain similar sub-folders from Audio to Video as described in previous paragraphs.

If the user prefers to use the mobile version of the TASK tool, please consult the details in Section 1.10.

1.1.2 Settings for PLAY700 App

Fig 1.3 shows a screenshot of the main menu window for the PLAY700 App where the reader can see a "gear' icon on the top right corner. Once this "gear" icon is tapped once, the "Settings" screen is shown as in Fig. 1.4.

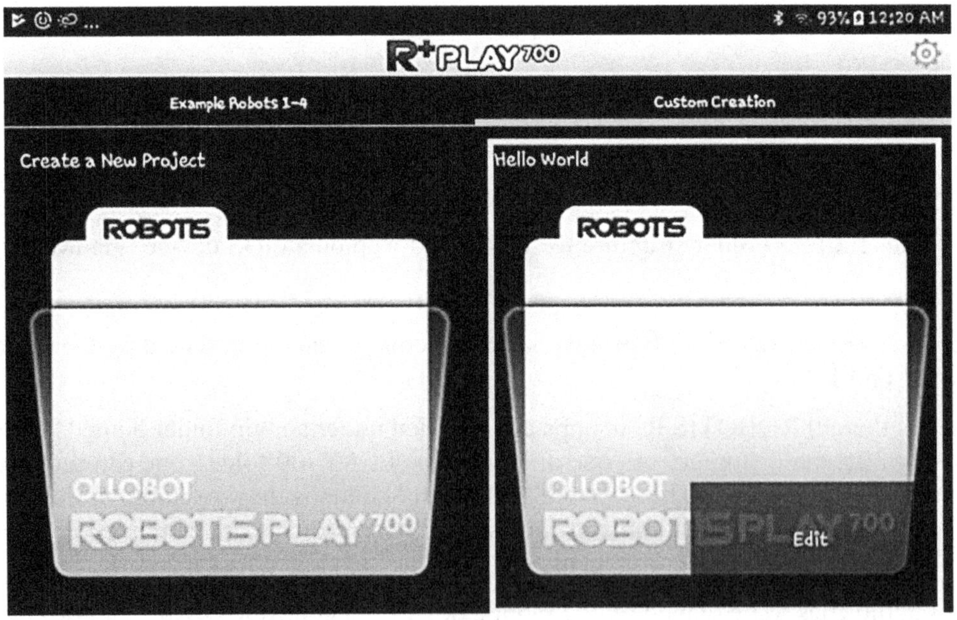

Fig. 1.3 Main Menu of PLAY700 app, opening on Custom Project "Hello World".

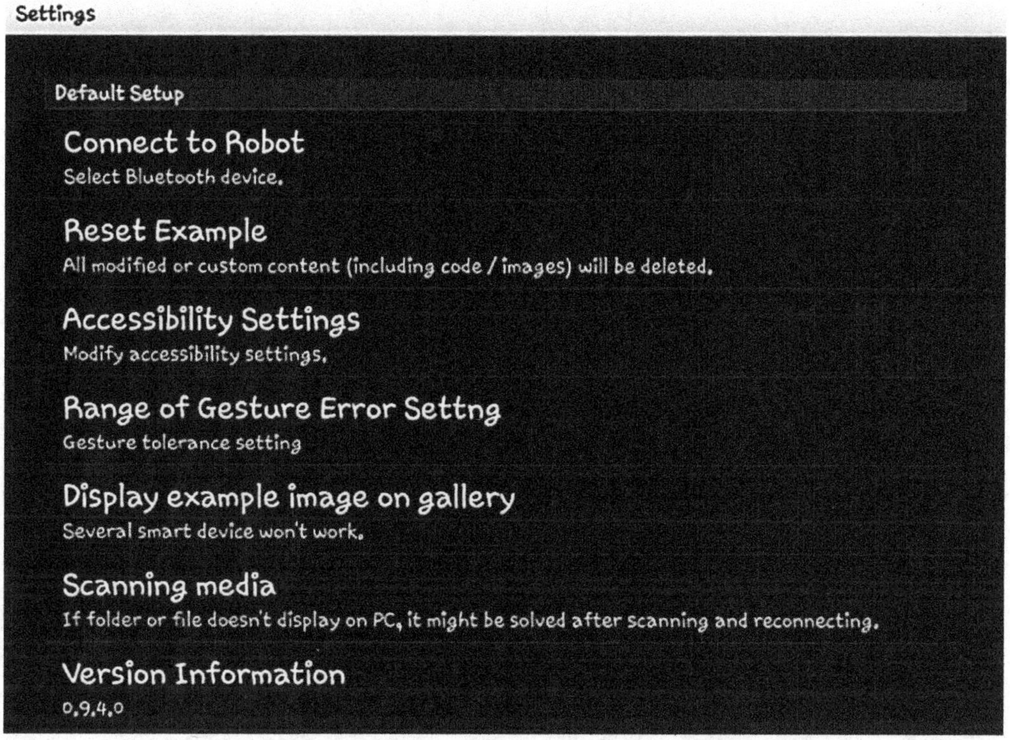

Fig. 1.4 App-level "Settings" for PLAY700 App.

The first item "Connect to Robot" allows the user to scan for new BT devices and pair them or to choose a specific BT device among those that were previously paired.

The second item "Reset example" can be used to reset all ROBOTIS example System projects to their originally installed states. This setting does not affect the Custom projects.

The third item "Accessibility Set-up" only pertains to Android devices where the user could set-reset many OS-level settings such as "Auto Rotate Screen" or "Screen Timeout". This "Accessibility" setting is not available for iOS devices.

The fourth item "Range of Gesture Error Setting" is used to set the accuracy of the gesture function (a number between 0 and 30 – available to both Android and iOS devices).

The fifth item "Display Example Image on Gallery" also pertains to Android devices only and does what its label implies.

The sixth item "Scanning Media" also pertains to Android devices only and removes the original "hidden" status of the folders and files inside the PLAY700 main folder so that Windows Explorer on a PC can "see" them.

1.1.3 Tools available for a PLAY700 Project

Going back to Fig. 1.3, when the user taps on the "Edit" button of a project, a screen as shown in Fig. 1.5 displays all the components/tools that can be used for this project. The current Android version (0.9.4.0) has all these tools operational, **but for the current iOS version (1.0.5), the following six tools are not yet functional**:

"Instrument" in the Multimedia group.
"Illumination" in the "Sensor" group.
"Received SMS", "Status Bar", "Vibration", "Application" in the "Other" group.

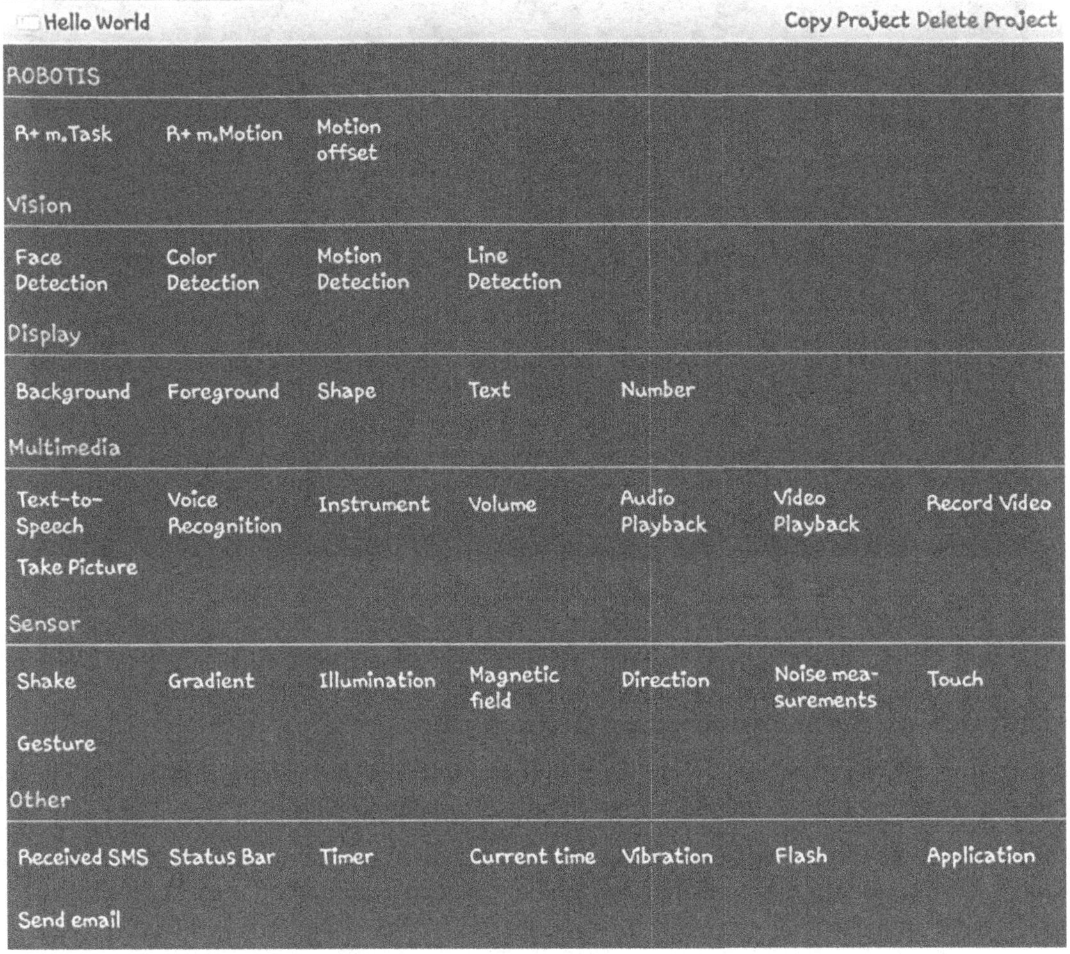

Fig. 1.5 Components/Tools of a SMART Project.

For example:

1) The "R+m.TASK" item allows the user to edit the TSKX files corresponding to this project.
2) The "Face Detection" item switches to the camera whereas the user can choose the front or back camera so that it can find a human face and then puts on a pair of sun glasses across the eyes, all in "real time".
3) The "Color Detection" item also kicks in the camera and shows a live image with a small centered region of interest where it would try to detect for 4 "colors": Black, Red, Green and Blue.

4) The "Display" section pertains to "Background" and "Foreground" images, and other items like "Text" and "Number". For example, the user can add and register Text Items if the user chooses the "Text" item in this "Display" section.
5) The "Multimedia" section can be used to record or play audio and video segments, and for text-to-speech and speech recognition tasks.
6) The "Sensor" section pertains to sensors embedded in the mobile device such as its Touch, Gesture, Gyro and Tilt sensors.
7) The "Other" section applies to "Messages" and other "Applications" that can be linked to this project.

1.2 Basic Usage of TASK with R+m.PLAY700

1.2.1 Two ways to access SMART DEVICE commands

Interestingly, the access to the SMART DEVICE commands for the CM-150 has a rather tortuous history which the interested reader can read at this web link https://www.cntrobotics.com/robotis-evolving-dynamixel-concept). At present, users have two options for using the SMART DEVICE commands:

1) **Using the TASK V.1 tool** which is part of the older RoboPlus V.1 Suite (http://www.robotis.us/roboplus1/). Fig. 1.6 illustrates the "SMART APP" menu in TASK V.1.

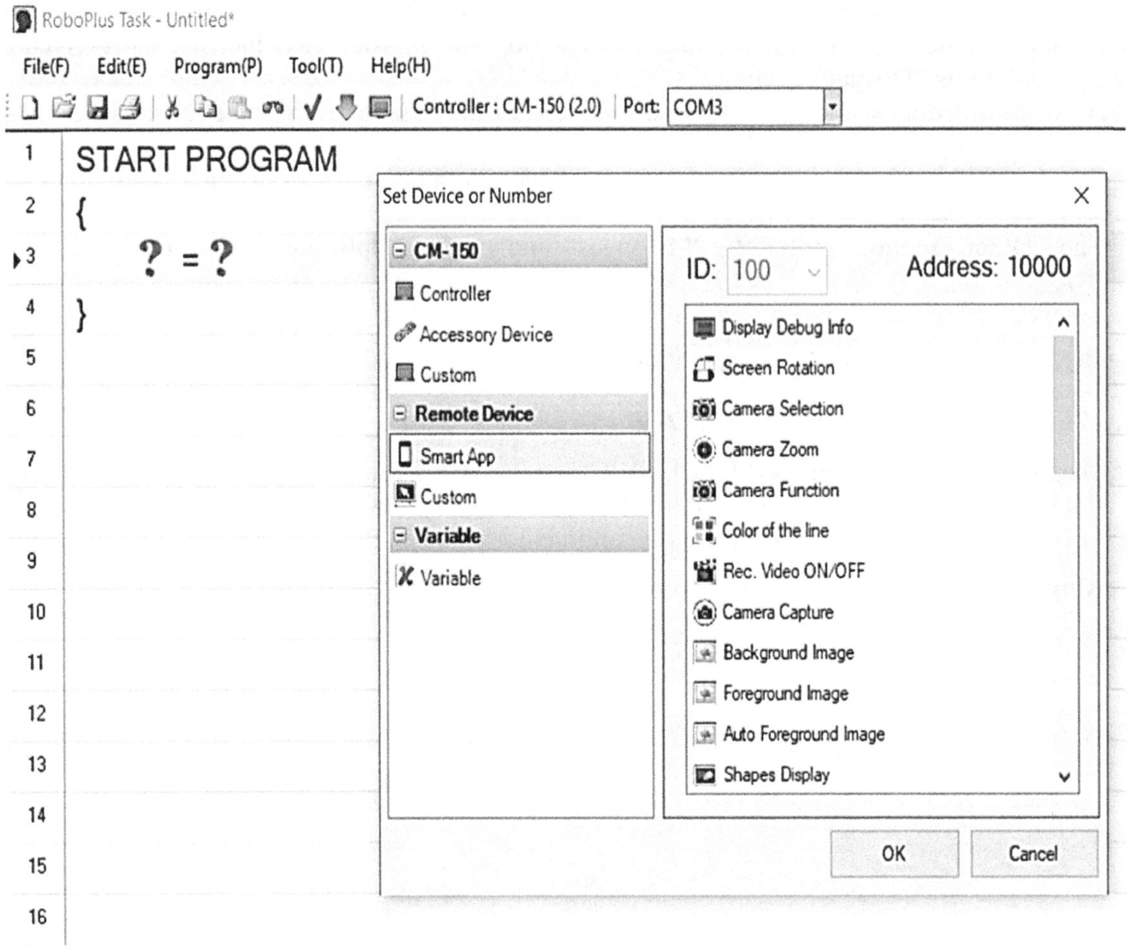

Fig. 1.6 "SMART APP" menu in TASK V.1 tool for CM-150.

The reader can see a partial view of an extensive set of mobile services available to this TASK tool in the right panel of Fig. 1.6. A generated TASK program is saved in the old TSK format and the compiled TSK file will interact perfectly fine with the R+m.PLAY700 App at run time. Video 1.1 documents the creation an example TSK program named "AD-HelloWorld.tsk" and its corresponding PLAY700 project named "Hello World". Video 1.1 also shows their run-time performance which essentially displays "Hello World!" for 5 seconds on a mobile device, whenever the user claps hands or whistles into it.

2) **Using the TASK V.2 tool** requires a more elaborate procedure because, starting with TASK V.2.1.4 (March 2017) and for unknown reasons, ROBOTIS removes access to the "SMART DEVICE" sub-menu for the CM-150's programming interface, however the actual SMART commands **are** still functional within the CM-150 firmware (*at least for now, i.e. April 2019*). Thus to use SMART commands with a CM-150, the user first has to temporarily change the "Programming Controller" field to one of the other Firmware 2.0 controllers (i.e. CM-50/200 or OpenCM-7.00/9.04, see Fig. 1.7 where CM-200 is used).

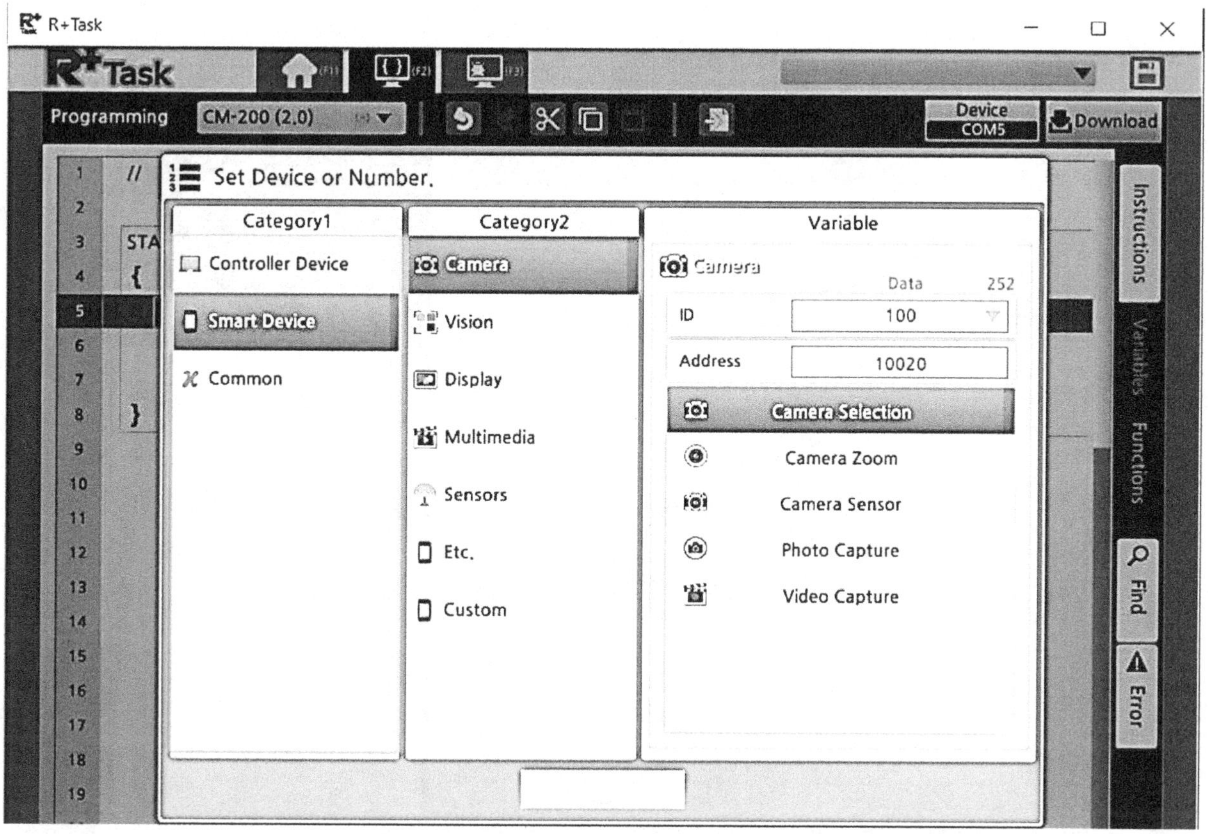

Fig. 1.7 "SMART DEVICE" menu shown in TASK V.2 tool for CM-200.

Then the user can access the SMART DEVICE commands as normal. However, the user needs to remember to switch back to the CM-150 as the "Programming Controller" to use the CM-150's specific functions and before compiling/downloading the resulting TASK code to a CM-150 robot – so a bit of annoyance to put up with until ROBOTIS changes its policy towards the CM-150 in the future (*hopefully*). The interesting point to note is that all those SMART DEVICE commands are retained in the user's TASK code once saved as a TSKX file! (see Fig. 1.8). Video 1.2 documents the conversion of "AD-HelloWorld.tsk" into "AD-HelloWorld.tskx" and its run-time performance.

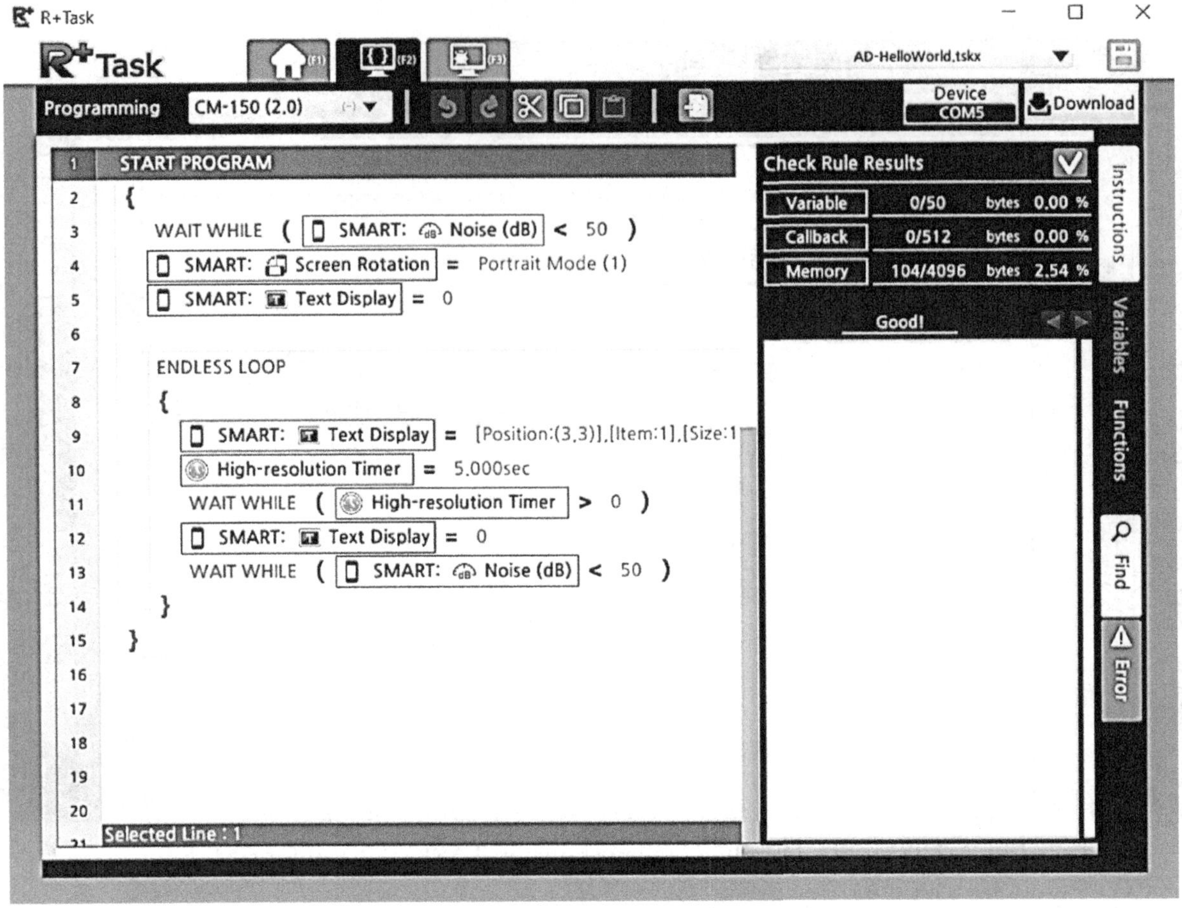

Fig. 1.8 "AD-HelloWorld.tskx" program in TASK V.2 tool for CM-150.

In the rest of this Chapter, we will be using the TASK V.2 tool as it has more functionalities than V.1, such as find and replace variables' names, syntax checking and debugging.

1.2.2 Organization of Display/Touch Zones on Mobile Device

As far as TASK is concerned, the Mobile Device's Display Screen is divided into 25 zones arranged in a 5x5 grid in either Portrait or Landscape mode (see Fig. 1.9).

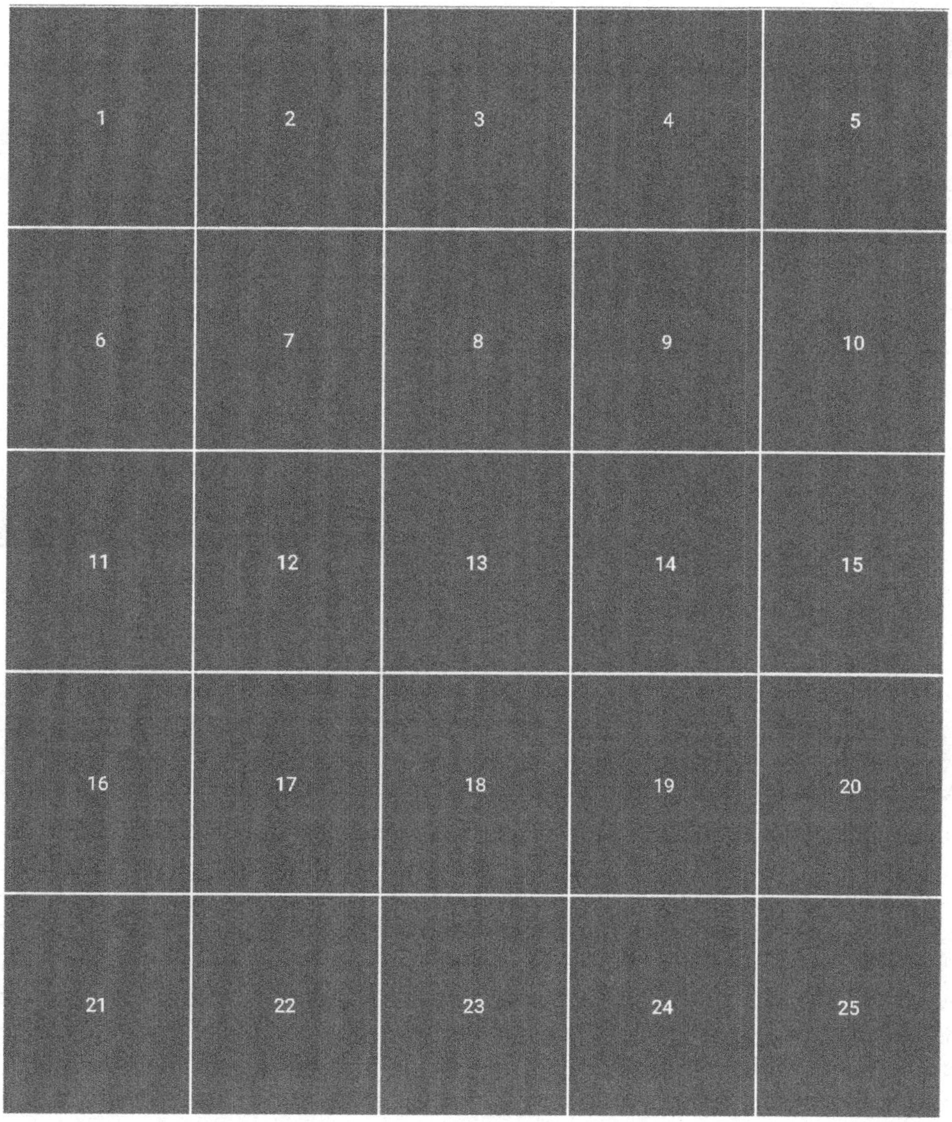

Fig. 1.9 Organization of the Display Screen on Mobile Device vis-à-vis TASK Tool.

When the Mobile Display is used as an Input Device, for example as a Touch Sensor (see more details in Section 1.3), the R+m.PLAY700 App reports a Zone Number between 1 and 25 (as shown in Fig. 1.9) to the TASK program at run time.

When the Mobile Display is used as an Output Device, for example for a Text item (Line 9 of Fig. 1.8 and see Fig. 1.10 for other functions such as "Shape" and "Number"), these Display Zones can be specified in two ways, depending on the programmer's needs, via a SMART CONSTANT Parameter which can have a fixed value or a variable value during run time:

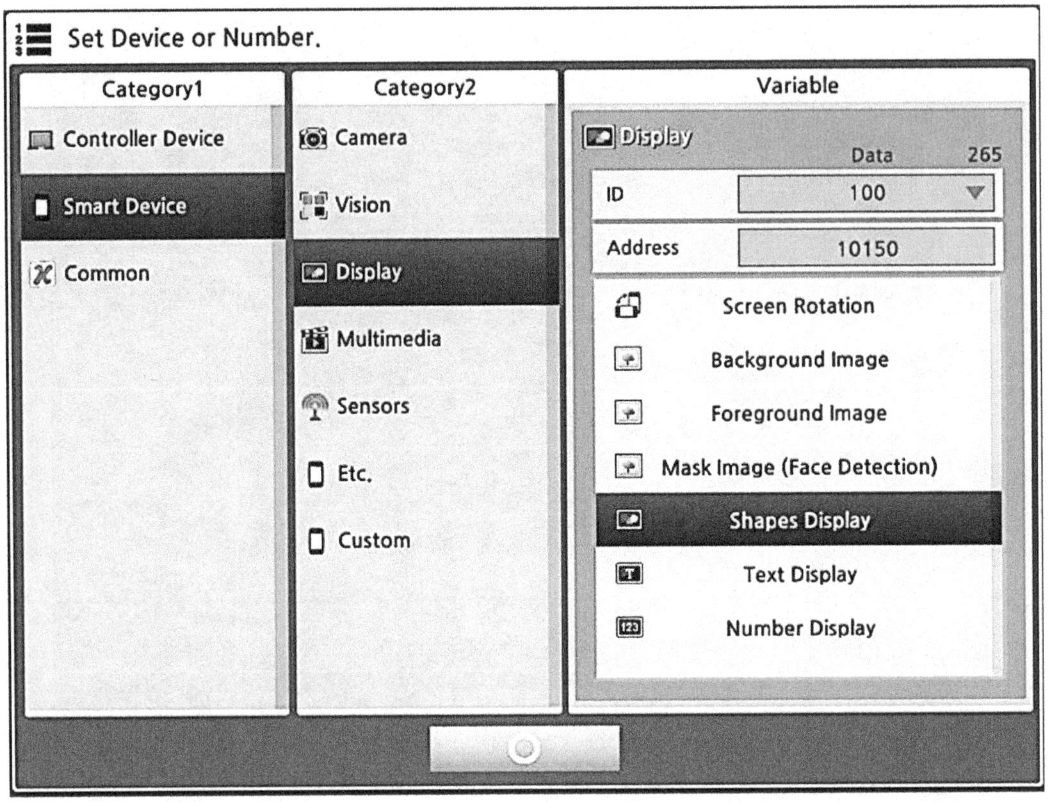

Fig. 1.10 Common Display Functions are for Shape, Text and Number in SMART DEVICE programming.

1) Fig. 1.11 illustrates the case where the parameters Position [3, 3] – i.e. Screen Center or Zone 13, Item No. [1], Size [100] and Color [White, i.e. 1] are all fixed values during run-time. This is the first way to specify the SMART CONSTANT Screen Position Parameter which is a 4-byte number (equal to "23331085" in this case - see Data field in red at the upper right corner of Fig. 1.11). The "SmartDeviceControlTable.PDF" file (at "address 10150") provides a detailed "formula" to compute the Screen Position Parameter (SPP) as follows:

SPP = **(Color * 16777216) + (Size * 65536) + (Item No * 256) + Position**

In this case, SPP = (1 * 16777216) + (100 * 65536) + (1 * 256) + 13 = **23331085**

In other words, "Position" forms the lowest byte of the 4 bytes used and "Item No" occupies the next higher byte (2nd byte). While "Size" is pushed into the 3rd byte, and "Color" is put into the highest 4th byte. **This formula applies to the display of Shape, Text and Number items**.

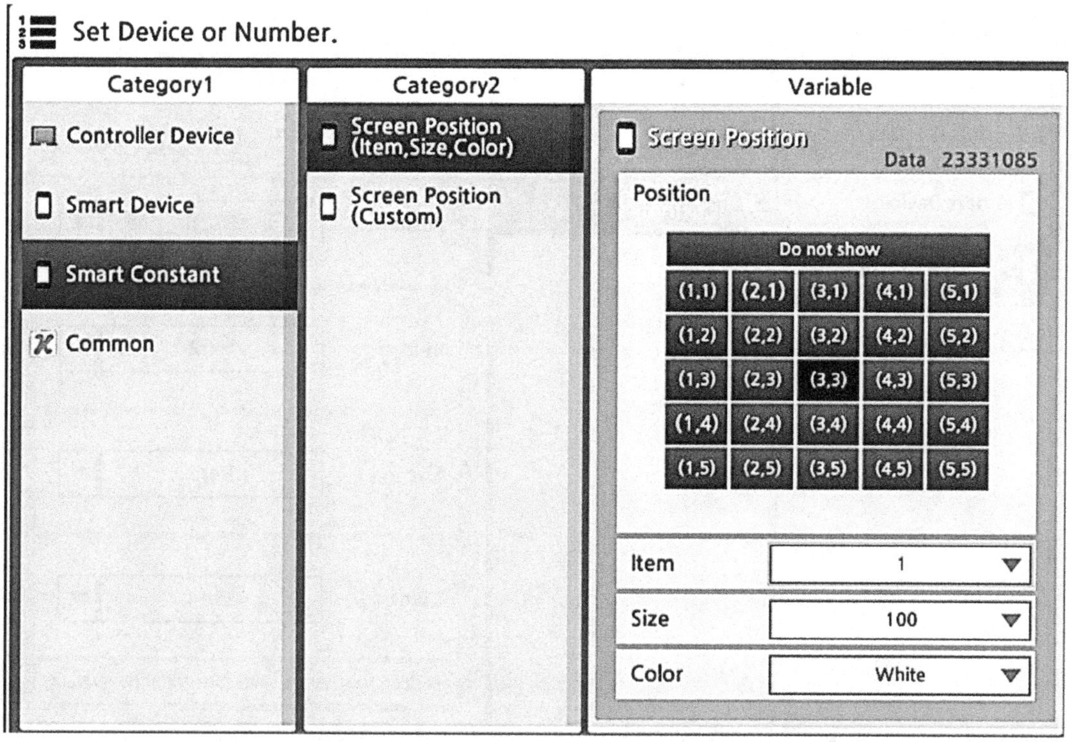

Fig. 1.11 Specification of a "fixed-value" Screen Position Parameter.

2) However, there may be times when the programmer needs to use a variable value for these parameters Position, Item No, Size or Color (i.e. variable arguments), then the programmer would need to use the "Custom" specification as illustrated in Fig. 1.12 where "needed" Parameters (from 1 to all 4) are defined by Variables named "Var1", "Var2", "Var3" and "Var4" respectively. Of course, these four "Var" Parameters would have to be defined with the correct values before their use here: for example, "Var1" should be set to "13" if the Screen Center is to be used for Display.

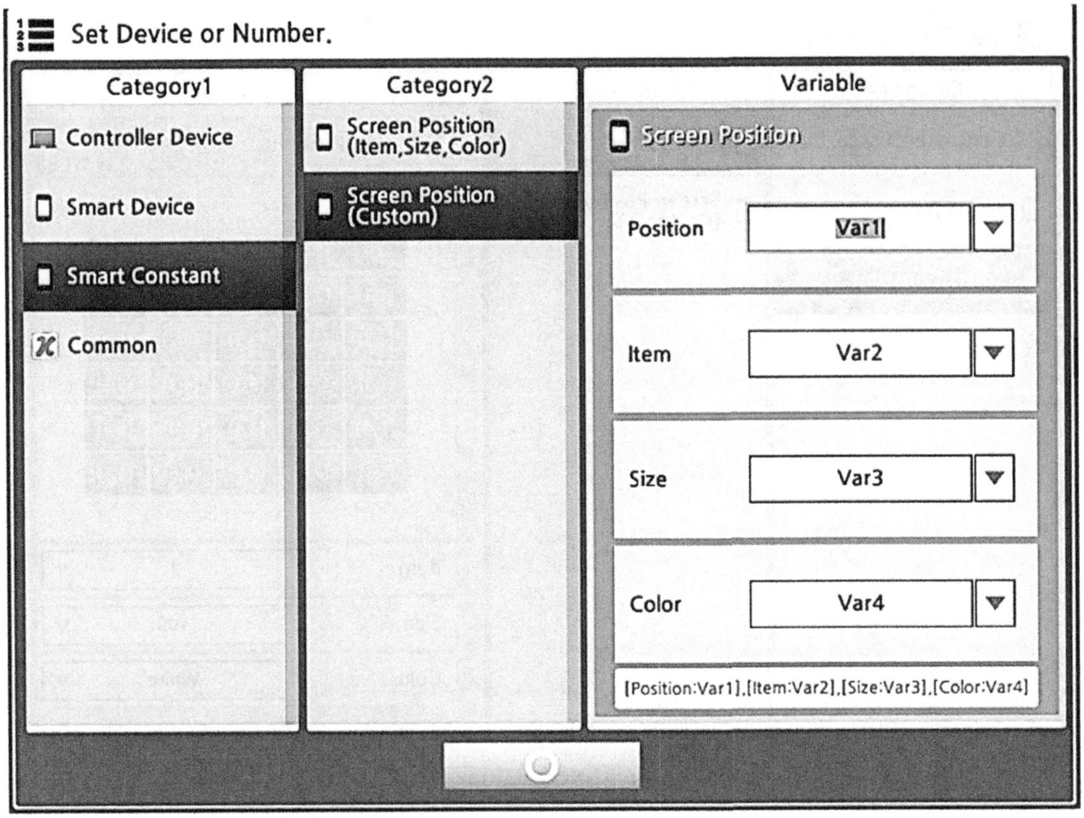

Fig. 1.12 Specification of a "variable-value" (i.e. Custom) Screen Position Parameter.

The program "AD-HelloWorld-Random.tskx" illustrates the use of a Custom Screen Position Parameter in conjunction with a Random Number (see Fig. 1.13):

1) Lines 5 to 8 initialize the values of Parameters PositionNo, ItemNo, SizeNo and ColorNo.

2) Line 9 waits on the user to make a loud enough noise to make the WAIT WHILE loop to exit to the next Line 10.

3) Line 10 puts the mobile display screen in Portrait Mode and Line 11 clears the screen.

```
5   PositionNo = 0
6   ItemNo = 1
7   SizeNo = 0
8   ColorNo = 0
9   WAIT WHILE ( SMART: Noise (dB) < 50 )
10  SMART: Screen Rotation = Portrait Mode (1)
11  SMART: Text Display = 0
12
13  ENDLESS LOOP
14  {
15      Random = 2 Random Number
16      PositionNo = Random % 26
17      SizeNo = Random % 100
18      ColorNo = Random % 10
19      IF ( PositionNo > 0 && ColorNo > 0 && SizeNo > 20 )
20      {
21          SMART: Text Display = [Position:PositionNo],[Item:ItemNo],[Size:SizeNo],[Color:ColorNo]
22          High-resolution Timer = 1.000sec
23          WAIT WHILE ( High-resolution Timer > 0 )
24          SMART: Text Display = 0
25          High-resolution Timer = 0.250sec
26          WAIT WHILE ( High-resolution Timer > 0 )
27      }
28  }
```

Fig. 1.13 Using Custom Screen Position in "AD-HelloWorld-Random.tstkx".

4) Then the Endless Loop (Lines 13-28) is entered whereas for each iteration, the following actions are performed:

a) A Random Number generated internally by the CM-150 is saved into Parameter Random (Line 15).

b) Next the Remainder Operator "%" is used to transform Parameter Random into an appropriate value for each of the Parameters PositionNo, SizeNo and ColorNo (Lines 16-18). The IF structure on Line 19 makes sure that various Parameters are in the proper ranges for reasonable display.

c) Line 21 combines the previous parameters into a Custom Screen Position to display Text Item 1 (i.e. "Hello World!") at a randomized screen position with randomized font sizes and colors.

d) Lines 22-23 allow the Text Display to stay on for 1 second, then Lines 24-26 clear the screen for 0.25 second.

Video 1.3 shows the run time performance of "AD-HelloWorld-Random.tskx". The reader should try to comment out Lines 24-26 to obtain a "collage" effect at run time.

If the reader is interested in a **manual computation** of the SMART CONSTANT described above, please read the program "AD-HelloWorld-Random-mc.tskx" (included in the example codes package).

1.3 Touch Areas and Shapes Display

Let's next look at a basic usage of the Touch Area feature provided with the PLAY700 App which can only monitor **up to 2** Touch Areas simultaneously. We'll also use the Shape Display feature whereas Shape #1 is a Circle, Shape #2 is a Square and Shape #3 is a Triangle (currently, PLAY700 does not allow any user-registered shape to be included in this list).

The program "AD-TouchMe.tskx" essentially monitors the 25 possible Touch Areas (see Fig. 1.9) where the user may have touched the mobile display with one or two fingers, and then it would draw either a circle or a square in those detected zones using randomized colors. Fig. 1.14 describes the algorithm used via an Endless Loop:

```
ENDLESS LOOP
{
    Touch1 = SMART: Touch Area 1
    Touch2 = SMART: Touch Area 2
    // Clearing screen display
    SMART: Shapes Display = 0
    IF ( Touch1 > 0 )
    {
        Random = Random Number
        ColorNo = Random % 10
        SMART: Shapes Display = [Position:Touch1],[Item:1],[Size:60],[Color:ColorNo]
        High-resolution Timer = 0.250sec
        WAIT WHILE ( High-resolution Timer > 0 )
    }
    IF ( Touch2 > 0 )
    {
        Random = Random Number
        ColorNo = Random % 10
        SMART: Shapes Display = [Position:Touch2],[Item:2],[Size:60],[Color:ColorNo]
        High-resolution Timer = 0.250sec
        WAIT WHILE ( High-resolution Timer > 0 )
    }
}
```

Fig. 1.14 Main Algorithm in program "AD-TouchMe.tskx".

1) Lines 13-14 describe how to use SMART DEVICE commands to access Touch Areas 1 and 2 where the user may have pressed on the mobile display. If no pressing is done, a zero value is saved into Parameters "Touch1" and "Touch2", otherwise a number between 1 and 25 will be returned from those SMART commands. The "first" user touch is saved into "Touch1" and the "second" touch is recorded into "Touch2".

2) Line 16 clears the screen.

3) Next, two parallel IFs are used, one for each Touch, as 1 or 2 touches can be performed by the user at run time. If "Touch1" is non-zero (line 17), a random number is drawn and used to set Parameter ColorNo with an appropriate value (Lines 19-20). Next, a Circle (i.e. Shape Item 1) is drawn at the location defined by "Touch1" in random color (line 21).

4) Next a 250 ms Time Delay is used (lines 22-23) to allow the user to see the displayed shape.

5) Lines 25-32 show a similar procedure used for Location "Touch2", but now a Square shape (i.e. Shape Item 2) with random color is used.

Video 1.4 shows the run time performance of "AD-TouchMe.tskx".

1.4 Text-to-Speech and Music Play

This musical project illustrates how to access the Musical Instruments available on the mobile device and how to activate the Text-to-Speech feature:

1) There are 128 Musical Instruments available via the PLAY700 App (from "Acoustic Grand Piano" [1] to "Gunshot" [128]). For each instrument, there are 10 Octave settings [1-10] and 12 Musical Scales or Notes [1-12]: e.g. Note 1 is "Do", while Note 3 is "Re", and so forth until Note 12 which is "Shi". A 3-byte SMART CONSTANT is associated with the Music Function where the lowest byte (Byte 1) contains the Note's value, while the next higher Byte (Byte 2) contains the Octave's value and Byte 3 contains the Instrument's value. The 4th byte of a typical SMART CONSTANT is not used for the Music Function (see Section 1.2.2) and needs to be set to zero.

2) The Text-to-Speech feature uses the same list of Text Items used for displaying them on the mobile screen (see Section 1.2.2) but activates a different SMART DEVICE function.

Fig. 1.15 shows the Initialization Event for the program "AD-MusicPlay.tskx":

1) Line 6 waits for the user to make a noise greater than 50 dB, essentially waiting for all the BT connections to be made on the mobile device side before starting the main algorithm.

2) Line 7 set the mobile screen into Portrait Mode and Line 8 clears the Display.

3) Lines 9-18 initialize the various parameters used in the program.

```
// User needs to make sure that robot is connected to mobil
WAIT WHILE ( SMART: Noise (dB) < 50 )
SMART: Screen Rotation = Portrait Mode (1)
SMART: Text Display = 0
InstrumentType = 0
Octave = 7
DoNote = 1
ReNote = 3
MiNote = 5
InstrumentValue1 = 0
InstrumentValue2 = 0
InstrumentValue3 = 0
InstrumentValueTemp1 = 0
InstrumentValueTemp2 = 0
```

Fig. 1.15 Initialization Event in program "AD-MusicPlay.tskx".

Fig. 1.16 shows Part 1 of the main Endless Loop for the program "AD-MusicPlay.tskx":

1) Line 23 starts a FOR LOOP using the Index "I" varying from 1 to 6 and incrementing by 1 each time.

2) Line 25 displays Text Item 7 (i.e. "Playing") and Line 26 displays Text Item I (whereas I varies from 1 to 6 as part of the implementation of the FOR LOOP).

3) Line 27 sends Text Item 7 to the Mobile Speech Engine to voice out "Playing" on the mobile device's speaker. Line 28 waits until this process is done before triggering the next Text-to-Speech event (i.e. Line 29). Lines 27-28 illustrate the "proper" use of the "Text-to-Speech" feature, that is to wait for the current Text Item to finish voicing out before starting on the next Text Item.

4) Lines 29-30 voices out the current type of Musical Instrument being played (i.e. Text Item I [1-6]).

```
20  ENDLESS LOOP
21  {
22      // Using Musical Instrument on Mobile Device for 3 Noted Do-Re-Mi - going through 6 different types of piano
23      LOOP FOR ( I = 1 ~ 6 )
24      {
25          SMART: Text Display = [Position:(3,2)].[Item:7].[Size:40].[Color:Green]
26          SMART: Text Display = [Position:(3,3)].[Item:I].[Size:40].[Color:Yellow]
27          SMART: Text to Speech (TTS) = Text Item 7
28          WAIT WHILE ( SMART: Text to Speech (TTS) != 0 )
29          SMART: Text to Speech (TTS) = I
30          WAIT WHILE ( SMART: Text to Speech (TTS) != 0 )
```

Fig. 1.16 Part 1 of Main Algorithm for program "AD-MusicPlay.tskx".

Fig. 1.17 shows Part 2 of the main Endless Loop for the program "AD-MusicPlay.tskx":

1) Line 32 save the current value of Index "I" to Parameter "InstrumentType" and Line 33 shifts this value 2 bytes to the left by multiplying it with the binary number "1 0000 0000 0000 0000" (i.e. to Byte 3 position, see Section 1.2.2 to review details about the structure of the 4-byte SMART CONSTANT). The decimal equivalent of the binary number "1 0000 0000 0000 0000" is "65536" (which was used in Section 1.2.2). This shifted "InstrumentType" value is then saved in Parameter "InstrumentValueTemp1".

2) Line 34 takes the current value of Octave [7] and shifts it by 1 byte to the left by multiplying it with "1 0000 0000" (i.e. to Byte 2 position). This shifted "Octave" value is then saved in Parameter "InstrumentValueTemp2".

3) Line 35 combines the current values of "InstrumentValueTemp1" and "InstrumentValueTemp2" and saves this result back to "InstrumentValueTemp1". Thus, Parameter "InstrumentValueTemp1" now has information about the Musical Instrument Type and the Octave level to be used. What is left is the Note to be set in Byte 1.

4) Lines 36-38 respectively add the Do-Re-Mi Notes into Parameter "InstrumentValueTemp1" and save the separate results into the respective final parameters "InstrumentValue1", "InstrumentValue2" and "InstrumentValue3".

5) Lines 39-40 then actually plays the Do Note on the "current" Instrument, next Lines 41-42 plays the Re Note, and finally Lines 43-44 plays the Mi Note.

```
            InstrumentType = 1
            InstrumentValueTemp1 = InstrumentType * 0000 0000 0000 0001 0000 0000 0000 0000
            InstrumentValueTemp2 = Octave * 0000 0000 0000 0000 0000 0001 0000 0000
            InstrumentValueTemp1 = InstrumentValueTemp1 + InstrumentValueTemp2
            InstrumentValue1 = InstrumentValueTemp1 + DoNote
            InstrumentValue2 = InstrumentValueTemp1 + ReNote
            InstrumentValue3 = InstrumentValueTemp1 + MiNote
            SMART: Play a musical instrument = InstrumentValue1
            WAIT WHILE ( SMART: Play a musical instrument > 0 )
            SMART: Play a musical instrument = InstrumentValue2
            WAIT WHILE ( SMART: Play a musical instrument > 0 )
            SMART: Play a musical instrument = InstrumentValue3
            WAIT WHILE ( SMART: Play a musical instrument > 0 )
        }
    }
```

Fig. 1.17 Part 2 of Main Algorithm for program "AD-MusicPlay.tskx".

Video 1.5 shows how the "Music" PLAY700 Project is set up on the mobile device and the run time performance of "AD-MusicPlay.tskx".

1.5 Remote Control using Touch Areas and Sound Effects

In this project, we'll use the Dump Truck (Robot 15) from the DREAM II School Set (see Fig. 1.18).

For this RC project, we'll again use the Touch Area feature but with two different user-interaction schemes:

1) For the standard Direction inputs such as Forward-Backward-Left-Right, the user needs to keep pressing on the selected Touch Area(s) for the robot to keep on performing the user-wanted maneuvers (i.e. "function activation upon pressing on Touch Area").

2) The user can press on two other Touch Areas to Increase or Decrease the robot's Speed independently of the current Direction input (subject of course to a maximum of two Touch Areas at run time), however the Speed change will occur only after the user lifts the finger off the selected Touch Area (i.e. "function activation upon release of Touch Area"). In other words, the user would need to "tap" a given Touch Area to make its related function to work.

We'll also implement two types of Sound Effects:

1) Using the CM-150 Buzzer to confirm a Speed Change.

2) Using the mobile device's Audio Playback service to play an audio clip of an engine running whenever the robot is in motion.

Fig. 1.18 Dump Truck robot from DREAM II School Set.

Fig. 1.19 describes the Initialization segment of the program "AD-TouchRC-DumpTruck.tskx":

1) Lines 5-7 initialize a Servo Motor on Port 3, and Lines 8-10 initialize Parameters Speed, TimeDelay1 and TimeDelay2.

2) Next, the mobile device screen is set to Portrait Mode (Line 12), clears the screen (Line 13) and waits for the user to make some noise louder than 50 dB (Line 15).

3) The "Directional" Text Items are set up with Lines 18-21, while the "Speed Control" Text Items are set up by Lines 23-24 and the "Platform Control" Text Items are set up by Lines 26-27. A PLAY700 project is also set on the mobile device with the following eight Text Items (in order): "Forward", "Backward", "Left", "Right", "Slow", "Fast", "Raise" and "Lower". An MP3 file named "CARDRIVING_KR.mp3" is also registered as Item 1 in the Audio Playback tool section (this file came from ROBOTIS with the PLAY700 Racing Car example).

```
PORT[3]:Servo Motor Drive Mode = TRUE (1)
PORT[3]:Servo Motor Speed = CCW:1000 (97.75%)
PORT[3]:Servo Motor Position = 512
Speed = 900
TimeDelay1 = 100
TimeDelay2 = 25
// Getting remote device ready using SMART commands
SMART: Screen Rotation = Portrait Mode (1)
SMART: Text Display = 0
// User needs to make sure that robot is connected to mobile device. User whistles loudy when ready.
WAIT WHILE ( SMART: Noise (dB) < 50 )

// Display Control Text FORWARD, BACKWARD, LEFT, RIGHT
SMART: Text Display = [Position:(3,2)],[Item:1],[Size:100],[Color:White]
SMART: Text Display = [Position:(3,4)],[Item:2],[Size:100],[Color:White]
SMART: Text Display = [Position:(2,3)],[Item:3],[Size:100],[Color:White]
SMART: Text Display = [Position:(4,3)],[Item:4],[Size:100],[Color:White]
// Display Speed Control Text SLOW or FAST
SMART: Text Display = [Position:(1,1)],[Item:5],[Size:75],[Color:Red]
SMART: Text Display = [Position:(5,1)],[Item:6],[Size:75],[Color:Red]
// Display Platform Control Text RAISE or LOWER
SMART: Text Display = [Position:(1,5)],[Item:7],[Size:75],[Color:Yellow]
SMART: Text Display = [Position:(5,5)],[Item:8],[Size:75],[Color:Yellow]
```

Fig. 1.19 Initialization Event for "AD-TouchRC-DumpTruck.tskx".

Fig. 1.20 describes Part 1 of the main algorithm implemented as an Endless Loop:

1) The values for the Detected Zones [1-25] for "Touch Area 1" and "Touch Area 2" are respectively saved in Parameters "Touch1" and "Touch2" (Lines 31-32).

2) When either Parameter Touch1 or Touch2 has a non-zero value (Line 33), this means that the user has pressed on the mobile screen somewhere, the algorithm goes on to check whether any Touch Area corresponding to the "Directional" zones has been confirmed using 4 PARALLEL IFs (Lines 35, 41, 47 and 53). In Thai (2018), the PARALLEL IFs structure is explained in detail from design goal to run-time performance, thus that information won't be repeated here. The gist of this structure is to allow the user to combine two "directional" inputs at the same time: for example, "Forward" and "Right" would make the robot make a wider right turn than if only "Right" was used.

3) Lines 37, 43, 49 and 55 are used to add a Sound Effect of a running engine out on the Play Audio 1 channel of the mobile device any time that the robot is set in motion. Please note that the PLAY700 App has a second Audio channel that can be used simultaneously with the first.

4) Finally, please note that the "directional" control of the robot is implemented with the "function activation upon pressing on Touch Area" scheme.

```
ENDLESS LOOP
{
    Touch1  =  SMART: Touch Area 1
    Touch2  =  SMART: Touch Area 2
    IF ( Touch1 != FALSE (0) || Touch2 != FALSE (0) )
    {
        IF ( Touch1 == 8 || Touch2 == 8 )
        {
            SMART: Play Audio 1  =  Audio Item 1
            CALL Forward
            CALL Delay1
        }
        IF ( Touch1 == 18 || Touch2 == 18 )
        {
            SMART: Play Audio 1  =  Audio Item 1
            CALL Backward
            CALL Delay1
        }
        IF ( Touch1 == 12 || Touch2 == 12 )
        {
            SMART: Play Audio 1  =  Audio Item 1
            CALL LeftTurn
            CALL Delay2
        }
        IF ( Touch1 == 14 || Touch2 == 14 )
        {
            SMART: Play Audio 1  =  Audio Item 1
            CALL RightTurn
            CALL Delay2
        }
```

Fig. 1.20 Part 1 of Main Algorithm for "AD-TouchRC-DumpTruck.tskx".

Fig. 1.21 describes Part 2 of the main algorithm:

```
60        IF ( Touch1 == 1 || Touch2 == 1 )
61        {
62            CALL Decrease_Speed
63        }
64        ELSE IF ( Touch1 == 5 || Touch2 == 5 )
65        {
66            CALL Increase_Speed
67        }
68
69        IF ( Touch1 == 21 || Touch2 == 21 )
70        {
71            CALL Raise_Platform
72        }
73        ELSE IF ( Touch1 == 25 || Touch2 == 25 )
74        {
75            CALL Lower_Platform
76        }
77    }
78    ELSE
79    {
80        SMART: Play Audio 1 = Not Used (0)
81        CALL Stop
82    }
```

Fig. 1.21 Part 2 of Main Algorithm for "AD-TouchRC-DumpTruck.tskx".

1) The "Decrease_Speed" or "Increase_Speed" alternatives are implemented together with an IF-ELSE-IF structure (Lines 60-67) as these two options are mutually exclusive logic-wise.

2) Similarly, the "Raise_Platform" and "Lower_Platform" alternatives are also implemented with another IF-ELSE-IF structure (Lines 69-76).

3) The final ELSE structure (Lines 78-82) is triggered when no Touch Area has been activated, whereas the Play Audio 1 Channel is shut down (Line 80) and the robot stops (Line 81).

The details for the Functions "Increase_Speed" and "Decrease_Speed" are shown in Fig. 1.22:

```
128 FUNCTION Increase_Speed
129 {
130     IF ( Speed < 1000 )
131     {
132         WAIT WHILE ( SMART: Touch Area 1 == 5 || SMART: Touch Area 2 == 5 )
133         Speed = Speed + 50
134         IF ( Speed >= 1000 )
135             Speed = 1000
136         Buzzer Timer = 0.2sec
137         Buzzer Index = Sol (10)
138     }
139 }
140
141 FUNCTION Decrease_Speed
142 {
143     IF ( Speed > 700 )
144     {
145         WAIT WHILE ( SMART: Touch Area 1 == 1 || SMART: Touch Area 2 == 1 )
146         Speed = Speed - 50
147         IF ( Speed <= 700 )
148             Speed = 700
149         Buzzer Timer = 0.2sec
150         Buzzer Index = Re (17)
151     }
152 }
```

Fig. 1.22 Function Definitions for "Increase_Speed" and "Decrease_Speed" in "AD-TouchRC-DumpTruck.tskx".

1) Function "Increase_Speed" is only "active" if (Speed < 1000) (Line 130), or in other words the robot's speed setting has an upper limit equals to 1000.

2) When (Speed < 1000), a WAIT WHILE loop is executed next (Line 132), implementing the scheme of "function activation upon release of Touch Area". Therefore, at run time the user needs to "press and release" (or "tap") the selected Touch Area representing "Fast" (zone 5) to affect a Speed Change with this scheme (*instead of "press and hold" like for "Directional" Touch Areas*).

3) Line 133 restricts each speed increase to 50 units each time and the IF structure in Lines 134-135 clamps Speed values to 1000 maximum.

4) Lines 136-137 implement an audible feedback each time that an actual speed increase has occurred by playing a Sol note for 0.2 second on the CM-150 buzzer.

5) Similarly, Function "Decrease_Speed" is only "active" if (Speed > 700) (Line 143), i.e. the robot has a lower speed limit equals to 700.

6) When (Speed > 700), a WAIT WHILE loop is executed next (Line 145), implementing the scheme of "function activation upon release of touch area". Therefore, at run time the user needs to "press and release" (or "tap") the selected Touch Area representing "Slow" (zone 1) to affect a Speed Decrease with this scheme.

7) Line 146 restricts each speed decrease to 50 units each time and the IF structure in Lines 147-148 clamps Speed values to a minimum of 700.

8) Lines 149-150 implement an audible feedback each time that an actual speed decrease has occurred by playing a Re note for 0.2 second on the CM-150 buzzer.

```
154 FUNCTION Raise_Platform
155 {
156     IF ( ServoPosition < 850 )
157     {
158         ServoPosition = ServoPosition + 50
159         PORT[3]:Servo Motor Position = ServoPosition
160         CALL Delay3
161     }
162 }
163
164 FUNCTION Lower_Platform
165 {
166     IF ( ServoPosition > 512 )
167     {
168         ServoPosition = ServoPosition - 50
169         PORT[3]:Servo Motor Position = ServoPosition
170         CALL Delay3
171     }
172 }
173
174 FUNCTION Delay3
175 {
176     Timer = 0.256sec
177     WAIT WHILE ( Timer > 0.000sec )
178 }
```

Fig. 1.23 Function Definitions for "Raise_Platform" and "Lower_Platform" in "AD-TouchRC-DumpTruck.tskx".

Fig. 1.23 has the details for Functions "Raise_Platform" and "Lower_Platform":

1) Lines 156-161 show that the "fully-raised" position of the Platform corresponds to Position 850 on Port 3 Servo Motor, and that each servo increment is by 50 units. A time delay of 0.256 second is implemented after each servo re-positioning to allow the user to stop the operation at any time during this platform raising maneuver.

2) Lines 166-171 show that the "fully-lowered" position of the Platform corresponds to Position 512 on Port 3 Servo Motor, and that each servo decrement is by 50 units. A time delay of 0.256 second is implemented after each servo re-positioning to allow the user to stop the operation at any time during this platform lowering operation.

Video 1.6 shows the run-time performance of the program "AD-TouchRC-DumpTruck.tskx".

1.6 Voice Control

For this project, we continue to use the Dump Truck as the demonstration platform (see Fig. 1.18) and allows for the same robot functionalities as described in the previous Section 1.5 (see Fig. 1.19), but this time these robot functions are activated by Voice Commands via the Speech Recognition tool of the PLAY700 App. Furthermore, because of the lengthy process of Speech Recognition (using the Web-based Google Speech Engine), each Voice Command will only trigger the corresponding robot function for a short time period, then the robot is reset to its full-stop state waiting for the next Voice Command.

Fig. 1.24 describes the Initialization segment of the program "AD-VoiceControl-WithDelays.tskx":

1) Lines 6-8 initialize a Servo Motor on Port 3, and Lines 9-14 initialize Parameters Speed, TimeDelay1, TimeDelay2, logical flag InvalidCommand and text item VoiceCommand.

2) Next, the mobile device screen is set to Portrait Mode (Line 16), clears the screen (Line 17) and waits for the user to make some noise louder than 50 dB (Line 19) which then triggers Line 20 setting the robot to its full-stop state.

3) Line 21 calls the Menu Function which is described in Fig. 1.25 which lists the TASK code on the left and the resulting mobile user interface on the right. The "Directional" Text Items are set up with Lines 145-149, while the "Speed Control" Text Items are set up by Lines 151-152 and the "Platform Control" Text Items are set up by Lines 154-155. A new Voice Control Text Item for "Talk" is also displayed at Position [3,5] (Line 157). A PLAY700 project is also needed to be set on the mobile device with the following eleven Text Items (in order from 1 to 11): "Forward", "Backward", "Left", "Right", "Slow", "Fast", "Raise", "Lower", "Talk", "Stop" and "Invalid Command".

```
// Voice Control of Dump Truck Maneuvers (CM-150)
// WITH time delays and stop for each command
// By C. N. Thai 3/20/2018
START PROGRAM
{
    PORT[3]:Servo Motor Drive Mode = TRUE (1)
    PORT[3]:Servo Motor Speed = CCW:1000 (97.75%)
    PORT[3]:Servo Motor Position = 512
    Speed = 900
    TimeDelay1 = 500
    TimeDelay2 = 250
    TimeDelay3 = 125
    InvalidCommand = FALSE (0)
    VoiceCommand = Text Item 10
    // Getting remote device ready using SMART commands
    SMART: Screen Rotation = Portrait Mode (1)
    SMART: Text Display = 0
    // User needs to make sure that robot is connected to mobile device. User whistles loudy when ready.
    WAIT WHILE ( SMART: Noise (dB) < 50 )
    CALL Stop
    CALL Menu
```

Fig. 1.24 Initialization Event for "AD-VoiceControl-WithDelays.tskx".

```
FUNCTION Menu
{
    SMART: Screen Rotation = Portrait Mode (1)
    SMART: Text Display = 0
    // Display Control Text FORWARD, BACKWARD, LEFT, RIGHT, STOP
    SMART: Text Display = [Position:(3,2)],[Item:1],[Size:60],[Color:White]
    SMART: Text Display = [Position:(3,4)],[Item:2],[Size:60],[Color:White]
    SMART: Text Display = [Position:(2,3)],[Item:3],[Size:60],[Color:White]
    SMART: Text Display = [Position:(4,3)],[Item:4],[Size:60],[Color:White]
    SMART: Text Display = [Position:(3,3)],[Item:10],[Size:60],[Color:Red]
    // Display Speed Control Text SLOW or FAST
    SMART: Text Display = [Position:(1,1)],[Item:5],[Size:60],[Color:Red]
    SMART: Text Display = [Position:(5,1)],[Item:6],[Size:60],[Color:Red]
    // Display Platform Control Text RAISE or LOWER
    SMART: Text Display = [Position:(1,5)],[Item:7],[Size:60],[Color:Yellow]
    SMART: Text Display = [Position:(5,5)],[Item:8],[Size:60],[Color:Yellow]
    // Display Voice Control Text TALK
    SMART: Text Display = [Position:(3,5)],[Item:9],[Size:60],[Color:Blue]
}
```

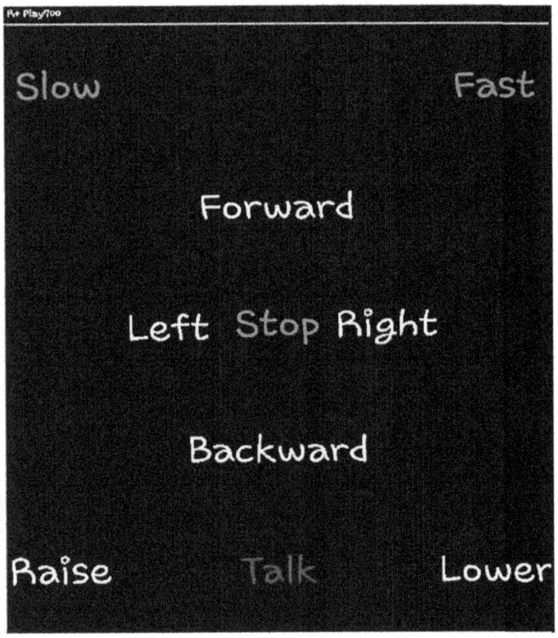

Fig. 1.25 Function "Menu" for "AD-VoiceControl-WithDelays.tskx".

Fig. 1.26 describes the main algorithm implemented as an Endless Loop:

1) The value for "Touch Area 1" is saved in Parameters "Touch1" (Line 25).

2) If Parameter Touch1 equals Position [3,5] (Line 26), then the structure defined by Lines 27 through 44 is entered, whereas:

a) Line 28 is used to wait for the user to lift the finger off the mobile device's screen.

b) Then Line 30 starts the Speech Recognition process (i.e. send a "1" to it) and the user would see the "Speak to Me" prompt appear on the mobile device, as shown on the right side of Fig. 1.26. The user then needs to voice out as soon as possible the possible verbal commands "Slow/Fast/Forward/Backward/Left/Right/Raise/Lower", as this process a has a built-in time-out period of about 5 seconds.

c) The Speech Recognition tool would then pick up the verbal input and try to match it up with its web database for the actual text as previous created in the PLAY700 App Project. This process is quite lengthy thus Line 32 is used to wait for its termination (i.e. when the parameter "SMART Speech Recognition" is set to 0 by the mobile device).

d) Next, the "Result of Speech Recognition" is saved into Parameter "VoiceCommand" (Line 33). If "VoiceCommand" is valid, i.e. not equal to -1, Function ProcessCommand is called (see details in Fig. 1.27) and the Parameter "Result of Speech Recognition" is cleared to 0 (Lines 34-38).

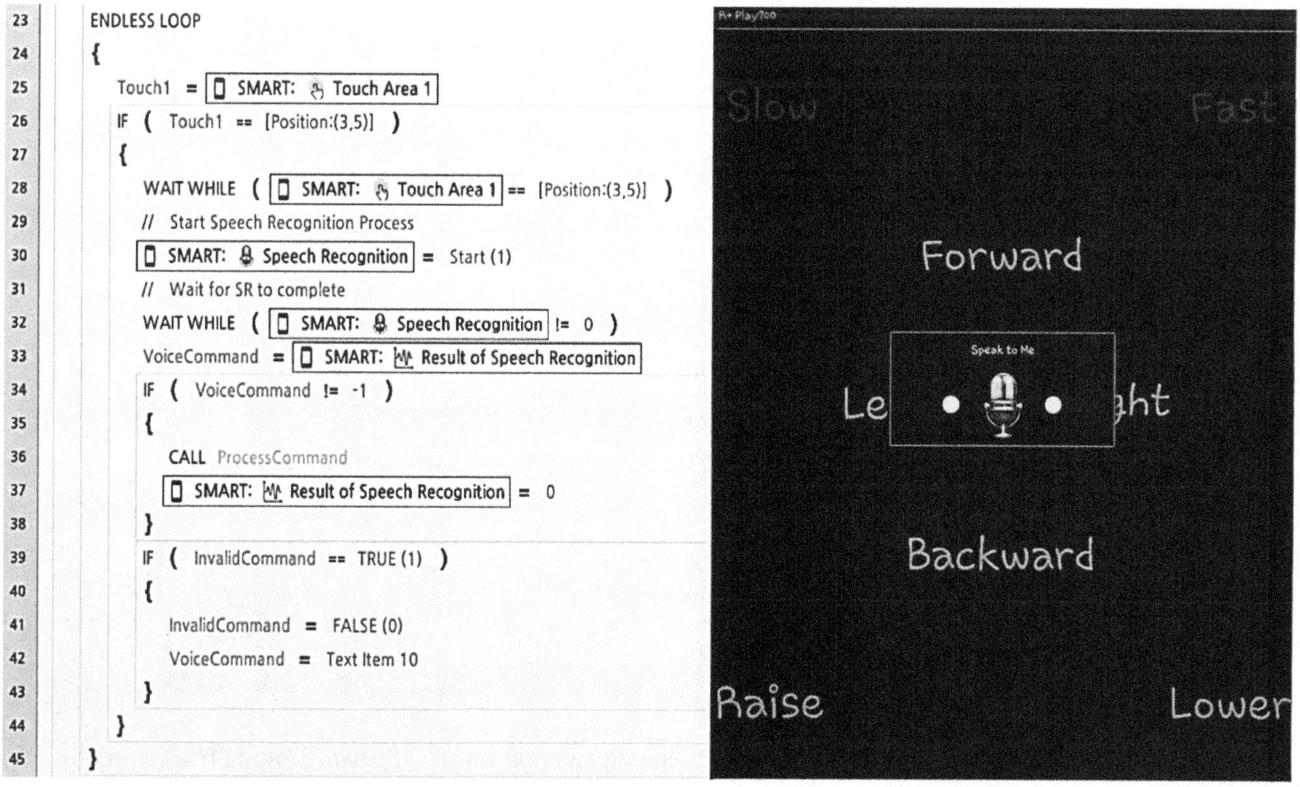

Fig. 1.26 Main Endless Loop for "AD-VoiceControl-WithDelays.tskx".

e) During the execution of Function "ProcessCommand" (see Fig. 1.27), Parameter "InvalidCommand" may be set to "TRUE" as part of an error catching procedure (Line 39), then Line 41 resets it to "FALSE", and Line 42 sets "VoiceCommand" to Text Item 10 (i.e. "Stop") and thus sets the robot in its default state.

3) Then the IF structure started out at Line 26 is finished and the Main Endless Loop is repeated once again at Line 25.

Figs. 1.27, 1.28, 1.29 and 1.30 describe the details of Function "ProcessCommand":

1) Fig. 1.27 describes the steps needed to process Text Item 1 (Forward) and Text Item 2 (Backwards), whereas:

a) When the user's verbal input is found to match with "Forward", the Function "Forward" is called, next Function "Delay1" delays the CM-150 Controller for 500 ms. Then the robot is stopped by calling Function "Stop" (Lines 164-169).

b) Similarly, when the user's verbal input is found to match with "Backward", the Function "Backward" is called, next Function "Delay1" delays the CM-150 Controller for 500 ms. Then the robot is stopped by calling Function "Stop" (Lines 170-175).

```
160 FUNCTION ProcessCommand
161 {
162     IF ( VoiceCommand != 0 )
163     {
164         IF ( VoiceCommand == Text Item 1 )
165         {
166             CALL Forward
167             CALL Delay1
168             CALL Stop
169         }
170         ELSE IF ( VoiceCommand == Text Item 2 )
171         {
172             CALL Backward
173             CALL Delay1
174             CALL Stop
175         }
```

Fig. 1.27 How Function "ProcessCommand" handles verbal inputs "Forward" and "Backwards" inside program "AD-VoiceControl-WithDelays.tskx".

2) Fig. 1.28 describes the steps needed to process Text Item 4 (Right) and Text Item 3 (Left), whereas:

a) When the user's verbal input is found to match with "Right" (Text Item 4), a "mild" Right Turn maneuver is executed by turning the robot "Right" for 250 ms (Lines 178-179), then "Forward" for another 250 ms (Lines 180-181). Next, this combination maneuver "Right-Forward" is repeated once more (Lines 182-185). Lastly, Function "Stop" is called to fulfill this "Right" command (Line 186).

```
176 ELSE IF ( VoiceCommand == Text Item 4 )
177 {
178     CALL RightTurn
179     CALL Delay2
180     CALL Forward
181     CALL Delay2
182     CALL RightTurn
183     CALL Delay2
184     CALL Forward
185     CALL Delay2
186     CALL Stop
187 }
188 ELSE IF ( VoiceCommand == Text Item 3 )
189 {
190     CALL LeftTurn
191     CALL Delay2
192     CALL Forward
193     CALL Delay2
194     CALL LeftTurn
195     CALL Delay2
196     CALL Forward
197     CALL Delay2
198     CALL Stop
199 }
```

Fig. 1.28 How Function "ProcessCommand" handles verbal inputs "Right" and "Left" inside program "AD-VoiceControl-WithDelays.tskx".

b) Similarly, when the user's verbal input is found to match with "Left" (Text Item 3), a "mild" Left Turn maneuver is executed by performing twice the combination maneuver "Left-Forward" (Lines 190-197). Finally, the Function "Stop" is called to fulfill this "Left" command (Line 198).

3) Fig. 1.29 describes the steps needed to process Text Item 10 (Stop), Text Item 5 (Slow), Text Item 6 (Fast), Text Item 7 (Raise) and Text Item 8 (Lower) whereas:

a) Line 200 shows that when the user's verbal input is found to match with "Stop" (Text Item 10), Function "Stop" is called to fulfill this "Stop" command (Line 202).

```
200 ELSE IF ( VoiceCommand == Text Item 10 )
201 {
202     CALL Stop
203 }
204 ELSE IF ( VoiceCommand == Text Item 5 )
205 {
206     CALL Decrease_Speed
207 }
208 ELSE IF ( VoiceCommand == Text Item 6 )
209 {
210     CALL Increase_Speed
211 }
212 ELSE IF ( VoiceCommand == Text Item 7 )
213 {
214     CALL Raise_Platform
215 }
216 ELSE IF ( VoiceCommand == Text Item 8 )
217 {
218     CALL Lower_Platform
219 }
```

Fig. 1.29 How Function "ProcessCommand" handles verbal inputs "Stop, "Slow", "Fast", "Raise" and "Lower" inside program "AD-VoiceControl-WithDelays.tskx".

b) When "Slow" is matched, Function "Decrease_Speed" is called (Line 206), and when "Fast" is matched, Function "Increase_Speed" is called (Line 210). The reader is encouraged to read on his/her own the TASK code segments for these two functions in the program "AD-VoiceControl-WithDelays.tskx" to verify that each speed increase or decrease was for 50 units, and that Parameter Speed was "clamped" to an upper limit of "1000" and a lower limit of "700", along with a different audio feedback for confirmation that an actual speed change had happened.

c) When "Raise" is matched, Function "Raise_Platform" is called (Line 214), and when "Lower" is matched, Function "Lower_Platform" is called (Line 218). The reader is also encouraged to read on his/her own the TASK code segments for these two functions in the program "AD-VoiceControl-WithDelays.tskx" to verify that each Servo Position increase or decrease was for 100 units, and that Parameter ServoPosition was "clamped" to an upper limit of "850" and a lower limit of "512".

4) Finally, Fig. 1.30 describes the last case that Function "ProcessCommand" needs to deal with, that is when Parameter "VoiceCommand" was set to "0" by the Speed Recognition tool, indicating that some sort of runtime error had happened (Lines 221-231):

```
220     }
221     ELSE
222     {
223         CALL Stop
224         InvalidCommand = TRUE (1)
225         SMART: Text Display = 0
226         SMART: Text to Speech (TTS) = Text Item 11
227         SMART: Text Display = [Position:(3,3)],[Item:11],[Size:100],[Color:Red]
228         WAIT WHILE ( SMART: Text to Speech (TTS) != 0 )
229         CALL Delay1
230         CALL Menu
231     }
232 }
```

Fig. 1.30 How Function "ProcessCommand" handles a runtime error reported by the Speech Recognition tool inside the program "AD-VoiceControl-WithDelays.tskx".

a) The first action is to stop the robot (Line 223), then the flag "InvalidCommand" is set to TRUE.

b) Next, Text Item 11, i.e. "Invalid Command", is displayed and voiced out on the mobile device to notify the user of this error situation for at least 500 ms via Function "Delay1" (Lines 225-229).

c) To wrap things up, the Function "Menu" is called to restore the original user interface as shown in Fig. 1.25 (Line 230).

Part 1 of Video 1.7 shows the run-time performance of the program "AD-VoiceControl-WithDelays.tskx", while its Part 2 shows the performance of the program "AD-VoiceControl-NoDelays.tskx" where the robot keeps on processing a previous valid command until it is changed by a new verbal command from the user. The reader is encouraged to learn through this last program on his/her own.

1.7 Bull Fight using NIR and Color Sensors with Audio Files

This Bull Fight project is a remake of a similar project described in Section 6.6 of Thai (2018) which only used the musical melodies from the built-in buzzer for sound effects. But this time around, we'll be using more realistic MP3 sound clips of a growling bull and of an "Ole" exclamation from the crowd attending a typical bullfight. These two sound clips come from the Sound Bible web site (http://soundbible.com/suggest.php?q=bull&x=0&y=0), courtesy of Mike Koenig and as a Creative Commons Attribution 3.0 credit.

Hardware wise, this Bull Robot (Fig. 1.31) illustrates a dual-range object detection scheme whereas:

Fig. 1.31 Bull Robot with Color Sensor facing forward.

- For the farther range, the built-in Left and Right NIR Sensors are used to detect the presence of the object and to make the robot move closer to the object. During these maneuvers, the PLAY700 App plays a growling-bull MP3 file out of Audio Channel 1 of the mobile device.

• When the Color Sensor on Port 4 gets in range, then it is used to determine whether the object has a Red color or not. If a Red color is found, the Bull robot would back up and then charge ahead while growling out of Audio Channel 1. At the end of this charge, the PLAY700 App plays the "Ole" MP3 file out of Audio Channel 2 of the mobile device. For any other color found, the robot just stops and plays Melody 20.

The resulting TASK code "AD-BullFight.tskx" has a Main Algorithm described as an Endless Loop in Fig. 1.32:

```
ENDLESS LOOP
{
    // Turns into the direction the object is detected. When both IR
    IF ( IR Left > 10 )
    {
        CALL Growl
        CALL Left
        IF ( IR Left >= 100 && IR Right >= 100 )
        {
            CALL Check_Color
        }
    }
    ELSE IF ( IR Right > 10 )
    {
        CALL Growl
        CALL Right
        IF ( IR Left >= 100 && IR Right >= 100 )
        {
            CALL Check_Color
        }
    }
    ELSE
    {
        CALL Stop
    }
}
```

Fig. 1.32 Main Algorithm for "AD-BullFight.tskx".

• For example, when the robot detects an object on its Left, it would growl and turn left to track it (lines 19-22). Functions "Growl" and "Ole" are simple SMART DEVICE statements to play Audio Items 1 and 2 out of Audio Channels 1 and 2 respectively (see Fig. 1.33).

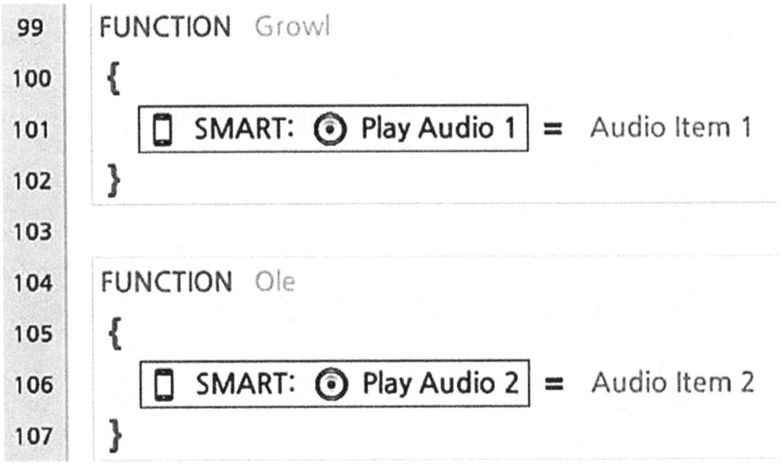

Fig. 1.33 Functions "Growl" and "Ole" used in "AD-BullFight.tskx".

• As the robot gets closer to the object, the Left and Right IR Sensors would increase in value, and when both are larger than 100, the robot checks the Color Sensor on Port 4 by calling Function Check_Color (Lines 23-26).

• A similar approach is used if the object is discovered on the robot's Right side (Lines 28-36).

• If no object is found, the robot stops (Lines 37-40).

Details for the Function "Check_Color" are shown in Fig. 1.34:

• First, Parameter CB_Color is obtained by using the special CALLBACK Function (http://support.robotis.com/en/software/roboplus/roboplus_task/programming/command/roboplus_task_cmd_callback.htm) which is activated independently of the main TASK program every 8 ms via the CM-150 hardware timer interrupt (Lines 125-129).

• Next, if (CB_Color == Red) happens to be TRUE (Line 111), Functions "Growl" and "Charge" are called (Lines 113-114), otherwise the robot stops and Melody 20 is played (Lines 118-121).

```
FUNCTION Check_Color
{
    IF ( CB_Color == Red Color (3) )
    {
        CALL Growl
        CALL Charge
    }
    ELSE
    {
        CALL Stop
        Buzzer Timer = Melody Time
        Buzzer Index = Melody No.20 (20)
        WAIT WHILE ( Buzzer Timer > 0.0sec )
    }
}

CALLBACK
{
    CB_Color = PORT[4]:Color Sensor
    // Print Screen with Line = CB_Color
}
```

Fig. 1.34 Functions “Check_Color” and CALLBACK used in “AD-BullFight.tskx”.

```
FUNCTION Charge
{
    // Moves backward then charge
    CALL Backward
    CALL Standby_0_25sec
    CALL Stop
    CALL Standby_0_25sec
    CALL Forward
    CALL Standby_0_75sec
    CALL Ole
    CALL Stop
}
```

Fig. 1.35 Function “Charge” used in “AD-BullFight.tskx”.

Details for the Function "Charge" are shown in Fig. 1.35:

- First, remember that the robot is "growling" at this point (Line 113 in Fig. 1.34) and so it rolls backwards for 0.25 second (Lines 47-48), then stops for another 0.25 second (Lines 49-50).
- Next, the robot charges forward for 0.75 second, then plays "Ole" and finally stops (Lines 51-54).

Video 1.8 shows the run-time performance of the program "AD-BullFight.tskx" which performs quite satisfactorily. The reader may remember from Section 6.7 of Thai (2018) that it was not possible to obtain a similar level of performance using the R+SCRATCH/SCRATCH 2 tool chain, as too much systemic delays occurred with this SCRATCH setup (see video at https://www.youtube.com/watch?v=ONxUAEUDkKE).

1.8 Camera Functions

Using the R+m.PLAY700 App, both rear and front cameras for a typical mobile device are accessible via SMART DEVICE commands inside a TASK program (please check program "AD-SelectCamera.tskx") along with picture and video capturing functions (see Fig. 1.36). Furthermore, some rudimentary image processing tools are also available through TASK programming:

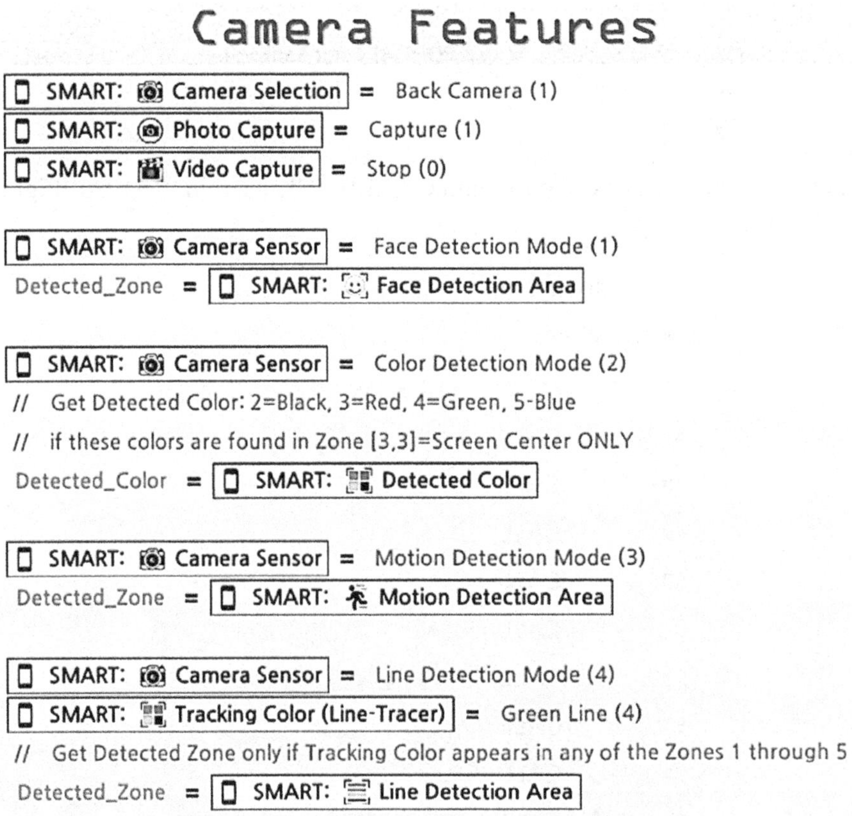

Fig. 1.36 R+m.PLAY700 Vision-related Functions available via TASK Programming.

1) In a "Face Detection" mode, the SMART "Face Detection Area" Function would return a number between 1 and 25 to the TASK program, representing the Zone Number where the "face" was detected (please review Fig. 1.9 for a map of these zones and try out the program "AD-FaceDetection.tskx")

2) In a "Color Detection" mode, only pixels within Zone 13 (i.e. Screen Center or Position [3,3]) are evaluated for their color values to see if they fit, on the average, "fixed" internal numerical RGB criteria for 4 possible color values: 2 for Black, 3 for Red, 4 for Green and 5 for Blue. Please try out program "AD-ColorSensing.tskx" and note that the App would classify a "Yellow" object as being "Red", so this App is not quite robust as users may expect.

3) In a "Motion Detection" mode, the SMART "Motion Detection Area" Function would return a number between 1 and 25 to the TASK program representing the Zone Number where "Motion" was detected (please try out the program "AD-MotionDetection.tskx").

4) In a "Line Detection" mode, the user needs to specify which color the App should be tracking for: 2 for Black, 3 for Red, 4 for Green and 5 for Blue. Furthermore, only the top 5 Screen Zones, i.e. Zone 1 through 5, are considered by the App, thus the SMART "Motion Detection Area" Function would return a number between 1 and 5 to the TASK program if the user-chosen color was found among those zones (please try out the program "AD-ColorLineDetection.tskx").

Video 1.9 illustrates the above 5 basic uses of the mobile camera.

1.8.1 Picture and Video Capture

This project "AD-CameraCapture.tskx" integrates the following SMART features into a more utilitarian application: text display, touch area, camera selection, picture/video capture, and activating an external App (i.e. Gallery) to review the recorded pictures and videos. The corresponding PLAY700 App project uses the following list of Text Items:

1 – Front Camera

2 – Back Camera

3 – Pictures

4 – Videos

5 – Selected

6 – Taking

7 – First Picture

8 – Second Picture

9 – Gallery

Fig. 1.37 describes the Main part of the program "AD-CameraCapture.tskx":

1) Lines 5 to 6 initialize Parameters "Camera" (Back or Front), RecordType (Pictures or Videos) and the Flag "Recording" (TRUE or FALSE).

2) The robot then waits for the user to make a noise louder than 50 (Line 9).

3) Next the mobile screen is set to Portrait mode (Line 11) and the Text Display overlay is cleared (Line 12).

```
3  START PROGRAM
4  {
5      Camera  =  0
6      RecordType  =  0
7      Recording  =  FALSE (0)
8      //  User needs to make sure that robot is connected to mobile device. User whistles loudy when ready.
9      WAIT WHILE  ( SMART: Noise (dB) <  50  )
10     //  Getting remote device ready using SMART commands
11     SMART: Screen Rotation  =  Portrait Mode (1)
12     SMART: Text Display  =  0
13     CALL  Menu
14
15     ENDLESS LOOP
16     {
17         IF  (  Camera  ==  0  ||  RecordType  ==  0  ||  Recording  ==  FALSE (0)  )
18         {
19             CALL  CheckTouch
20         }
21         IF  (  Recording  ==  TRUE (1)  )
22         {
23             CALL  Capture
24         }
25     }
26 }
```

Fig. 1.37 Main Part of Program "AD-CameraCapture.tskx".

4) Line 13 calls Function "Menu" to be described in Fig. 1.38.

5) The Main Endless Loop (Lines 15-25) is then entered and two parallel IF structures are executed for each loop iteration:

- The first IF structure (Lines 17-20) checks to see if the 3 parameters "Camera", "RecordType" and "Recording" are still at their initial values (i.e. "0" and "FALSE"), meaning that the user has not finished making his or her choices to whether use the Back or Front Camera (i.e. "Camera" set to "1" or "2"), or to whether take capture Pictures or Videos (i.e. "RecordType" set to "1" or "2"), or that it is time to execute the Capturing process (i.e. "Recording" set to TRUE). If any of these 3 parameters is still at its initial value, then Function "CheckTouch" is called (described in Figs. 1.39 to 1.43).

- The second IF structure (Lines 21-24) checks to see if "Recording" is set to "TRUE". If "Recording is still set at "FALSE", the Endless Loop just continues. If "Recording" is set to "TRUE" then Function Capture is called (to be described in Figs. 1.44 and 1.45).

Fig. 1.38 describes the Menu (i.e. User Interface) for this project which consists of 5 Text Display and 1 Shape Display commands.

```
FUNCTION Menu
{
    SMART: Text Display = 0
    SMART: Text Display = [Position:(2,1)].[Item:1].[Size:40].[Color:White]
    SMART: Text Display = [Position:(2,2)].[Item:2].[Size:40].[Color:White]
    SMART: Text Display = [Position:(5,1)].[Item:3].[Size:40].[Color:White]
    SMART: Text Display = [Position:(5,2)].[Item:4].[Size:40].[Color:White]
    SMART: Text Display = [Position:(5,5)].[Item:9].[Size:40].[Color:White]
    SMART: Shapes Display = [Position:(3,5)].[Item:1].[Size:40].[Color:Green]
}
```

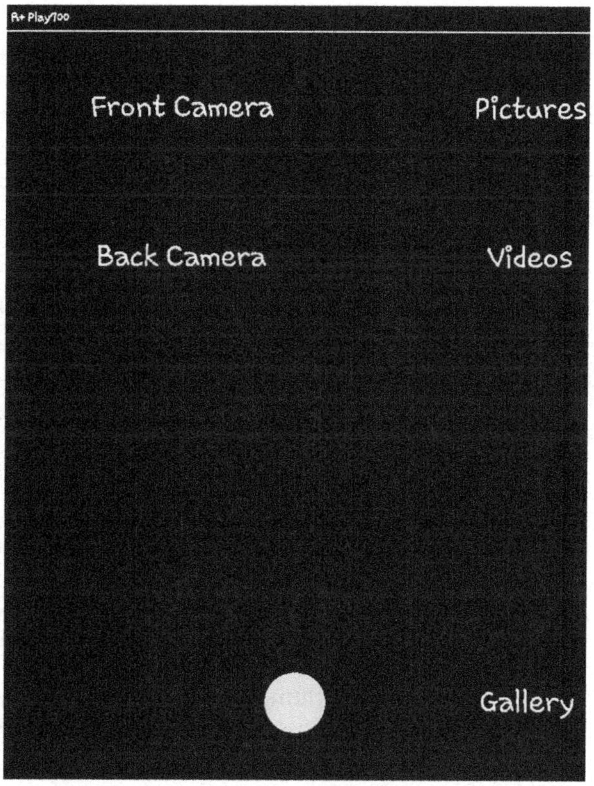

Fig. 1.38 Source Code for Function "Menu" and how it looks at runtime.

Figs 1.39 to 1.43 describe the lengthy Function "CheckTouch" which serves as the User Interface for this project:

1) In Fig. 1.39, first "Touch Area 1" is saved into Parameter Touch 1 (Line 41) and if (Touch1 == 0), meaning that the user did not tap on the mobile screen yet, then the Function "CheckTouch" is exited with no further action.

2) However, if the user has pressed somewhere on the mobile screen (Touch1 != 0), then a WAIT WHILE (Line 44) is used to wait for the user to lift the finger off the screen. Next, a lengthy IF-ELSE-IF structure (Lines 45 to 104) is used to determine the proper procedure to use depending on where the user taps on the mobile screen.

```
39 FUNCTION CheckTouch
40 {
41     Touch1 = SMART: Touch Area 1
42     IF ( Touch1 != 0 )
43     {
44         WAIT WHILE ( SMART: Touch Area 1 != 0 )
45         IF ( Touch1 == [Position:(1,1)] || Touch1 == [Position:(2,1)] )
46         {
47             Camera = 2
48             SMART: Text Display = [Position:(2,1)].[Item:1].[Size:40].[Color:Red]
49             SMART: Text Display = [Position:(3,3)].[Item:1].[Size:40].[Color:Yellow]
50             SMART: Text to Speech (TTS) = Text Item 1
51             WAIT WHILE ( SMART: Text to Speech (TTS) != 0 )
52             SMART: Text Display = [Position:(3,3)].[Item:5].[Size:40].[Color:Yellow]
53             SMART: Text to Speech (TTS) = Text Item 5
54             WAIT WHILE ( SMART: Text to Speech (TTS) != 0 )
55             SMART: Text Display = [Position:(3,3)].[Item:0].[Size:40].[Color:Black]
56         }
```

Fig. 1.39 Part 1 of Function "CheckTouch" in "AD-CameraCapture.tskx".

3) Lines 45-56 show that if "Front Camera" is chosen (i.e. Position (1,1) or (2,1)), then "Camera" is set to 2 and the "original" Text Item 1 "Front Camera" is over-written in Red at its original Position (2,1) (Line 48), also a second Text Item 1 is written in Yellow at Position (3,3) (Line 49) and verbalized with Lines 50-51.

4) Next the Text Item 5 "Selected" is displayed in Yellow at Position (3,3) and verbalized (Lines 52-54). Finally, Text Item 0 (i.e. nothing) is written to Position (3,3) in Black, essentially clearing out Display Zone (3,3) (Line 55).

5) Fig. 1.40 describes the tasks to use when Position (1,2) or (2,2) is touched, meaning that the user chose "Back Camera", so "Camera" is set to 1 (Line 59).

6) Next, the original Text Item 2 "Back Camera" is overwritten in Red by Line 60, and it is additionally displayed in Yellow at Position (3,3) and verbalized (Lines 61-63).

7) Next the Text Item 5 "Selected" is displayed in Yellow at Position (3,3) and verbalized (Lines 64-66). Lastly, Text Item 0 is written to Position (3,3) in Black, essentially clearing out Display Zone (3,3) (Line 67).

```
57 ELSE IF ( Touch1 == [Position:(1,2)] || Touch1 == [Position:(2,2)] )
58 {
59   Camera = 1
60   SMART: Text Display = [Position:(2,2)],[Item:2],[Size:40],[Color:Red]
61   SMART: Text Display = [Position:(3,3)],[Item:2],[Size:40],[Color:Yellow]
62   SMART: Text to Speech (TTS) = Text Item 2
63   WAIT WHILE ( SMART: Text to Speech (TTS) != 0 )
64   SMART: Text Display = [Position:(3,3)],[Item:5],[Size:40],[Color:Yellow]
65   SMART: Text to Speech (TTS) = Text Item 5
66   WAIT WHILE ( SMART: Text to Speech (TTS) != 0 )
67   SMART: Text Display = [Position:(3,3)],[Item:0],[Size:40],[Color:Black]
68 }
```

Fig. 1.40 Part 2 of Function "CheckTouch" in "AD-CameraCapture.tskx".

8) Fig. 1.41 describes similar tasks as before to use when Position (5,1) is touched, meaning that the user chose "Pictures", so "RecordType" is set to 1 (Line 71).

9) Next, the original Text Item 3 "Pictures" is overwritten in Red by Line 72, and it is additionally displayed in Yellow at Position (3,3) and verbalized (Lines 73-75).

10) Then, Text Item 5 "Selected" is displayed in Yellow at Position (3,3) and verbalized (Lines 76-78). Lastly, Text Item 0 is written to Position (3,3) in Black to clear out Display Zone (3,3) (Line 79).

```
69 ELSE IF ( Touch1 == [Position:(5,1)] )
70 {
71   RecordType = 1
72   SMART: Text Display = [Position:(5,1)],[Item:3],[Size:40],[Color:Red]
73   SMART: Text Display = [Position:(3,3)],[Item:3],[Size:40],[Color:Yellow]
74   SMART: Text to Speech (TTS) = Text Item 3
75   WAIT WHILE ( SMART: Text to Speech (TTS) != 0 )
76   SMART: Text Display = [Position:(3,3)],[Item:5],[Size:40],[Color:Yellow]
77   SMART: Text to Speech (TTS) = Text Item 5
78   WAIT WHILE ( SMART: Text to Speech (TTS) != 0 )
79   SMART: Text Display = [Position:(3,3)],[Item:0],[Size:40],[Color:Black]
80 }
```

Fig. 1.41 Part 3 of Function "CheckTouch" in "AD-CameraCapture.tskx".

11) Fig. 1.42 describes similar tasks as before to use when Position (5,2) is touched, meaning that the user chose "Videos", so "RecordType" is set to 2 (Line 83).

12) Next, the original Text Item 3 "Videos" is overwritten in Red by Line 84, and it is additionally displayed in Yellow at Position (3,3) and verbalized (Lines 85-87).

13) Then, Text Item 5 "Selected" is displayed in Yellow at Position (3,3) and verbalized (Lines 88-90). Lastly, Text Item 0 is written to Position (3,3) in Black to clear out Display Zone (3,3) (Line 91).

```
81  ELSE IF ( Touch1 == [Position:(5,2)] )
82  {
83    RecordType = 2
84    SMART: Text Display = [Position:(5,2)],[Item:4],[Size:40],[Color:Red]
85    SMART: Text Display = [Position:(3,3)],[Item:4],[Size:40],[Color:Yellow]
86    SMART: Text to Speech (TTS) = Text Item 4
87    WAIT WHILE ( SMART: Text to Speech (TTS) != 0 )
88    SMART: Text Display = [Position:(3,3)],[Item:5],[Size:40],[Color:Yellow]
89    SMART: Text to Speech (TTS) = Text Item 5
90    WAIT WHILE ( SMART: Text to Speech (TTS) != 0 )
91    SMART: Text Display = [Position:(3,3)],[Item:0],[Size:40],[Color:Black]
92  }
```

Fig. 1.42 Part 4 of Function "CheckTouch" in "AD-CameraCapture.tskx".

14) The first part of Fig. 1.43 (Lines 93-97) describes tasks to be done when the Green Circle at Position (3,5) is pressed, meaning that the user is ready to record "Pictures" or "Videos", so "Recording" is set to TRUE (Line 95), and the Circle at Position (3,5) is now painted Red (Line 96).

```
93   ELSE IF ( Touch1 == [Position:(3,5)] )
94   {
95     Recording = TRUE (1)
96     SMART: Shapes Display = [Position:(3,5)],[Item:2],[Size:40],[Color:Red]
97   }
98   ELSE IF ( Touch1 == [Position:(5,5)] )
99   {
100    SMART: Run the App. = App. Item 1
101    High-resolution Timer = 5.000sec
102    WAIT WHILE ( High-resolution Timer > 0.000sec )
103  }
```

Fig. 1.43 Part 5 of Function "CheckTouch" in "AD-CameraCapture.tskx".

15) The second part of Fig. 1.43 (Lines 98-103) describes tasks to be done when Position (5,5) is touched, meaning that the user is ready to jump to the App "Gallery" to review recorded pictures or videos, so App Item 1 is written to the SMART "Run the App" function (Line 100), and the Controller does nothing for 5 seconds (Lines 101-102) to let the user review saved recordings.

When Parameter Recording is set to TRUE, Function "Capture' is activated and Fig. 1.44 shows Part 1 of this Function where "Pictures" are recorded twice: first with No Zoom (Line 112) then secondly with Zoom set to 30 (Line 118):

1) First, the previously chosen camera is selected by Line 109.

2) If "Pictures" was chosen (RecordType == 1), then the chosen camera Zoom Level is set to zero (Line 112) and the first snapshot is taken (Line 113).

3) Text Item 6 ("Taking") is verbalized (Lines 114-115), then Text Item 7 ("First Picture" is verbalized (Lines 116-117).

```
107  FUNCTION Capture
108  {
109      SMART: Camera Selection = Camera
110      IF ( RecordType == 1 )
111      {
112          SMART: Camera Zoom = 0
113          SMART: Photo Capture = Capture (1)
114          SMART: Text to Speech (TTS) = Text Item 6
115          WAIT WHILE ( SMART: Text to Speech (TTS) != 0 )
116          SMART: Text to Speech (TTS) = Text Item 7
117          WAIT WHILE ( SMART: Text to Speech (TTS) != 0 )
118          SMART: Camera Zoom = 30
119          SMART: Photo Capture = Capture (1)
120          SMART: Text to Speech (TTS) = Text Item 6
121          WAIT WHILE ( SMART: Text to Speech (TTS) != 0 )
122          SMART: Text to Speech (TTS) = Text Item 8
123          WAIT WHILE ( SMART: Text to Speech (TTS) != 0 )
124          SMART: Camera Zoom = 0
125          SMART: Photo Capture = Stop (0)
126      }
```

Fig. 1.44 Part 1 of Function "Capture" in "AD-CameraCapture.tskx".

4) For the second snapshot, Zoom is set to 30 (Line 118) and similar tasks are followed to capture the second picture (Line 119) along with the appropriate verbalizations (Lines 120-123).

5) Finally, the Camera Zoom is reset to zero (Line 124), and the Photo Camera process is stopped (Line 125).

Fig. 1.45 shows Part 2 of the Function "Capture" where "Videos" are recorded:

1) Zoom is reset to zero (Line 129), and the Video Recording process is started by writing a "1" to the SMART Video Capture function (Line 130).

2) A waiting period of 2 seconds (Lines 131-136) is then used (the user can change this time constant to suit his or her needs). During this wait, Text Item 6 "Taking" and Text Item 4 "Videos" are displayed on the mobile device's screen (Lines 134-135).

```
127      ELSE IF ( RecordType == 2 )
128      {
129          SMART: Camera Zoom = 0
130          SMART: Video Capture = Recode (1)
131          High-resolution Timer = 2.000sec
132          LOOP WHILE ( High-resolution Timer > 0.000sec )
133          {
134              SMART: Text Display = [Position:(3,3)],[Item:6],[Size:40],[Color:White]
135              SMART: Text Display = [Position:(3,4)],[Item:4],[Size:40],[Color:White]
136          }
137          SMART: Video Capture = Stop (0)
138      }
139      Camera = 0
140      RecordType = 0
141      Recording = FALSE (0)
142      CALL Menu
143  }
```

Fig. 1.45 Part 2 of Function "Capture" in "AD-CameraCapture.tskx".

3) Line 137 is used to stop the video recording process, and we come to the end of the Function "Capture" where the three Parameters "Camera", "RecordType" and "Recording" are set to their initial values (Lines 139-141). Lastly the initial Menu is redisplayed (Line 142).

Video 1.10 shows the runtime performance of the program "AD-CameraCapture.tskx".

1.8.2 Color Dowel Alert

This project "AD-ColorDowelAlert.tskx" showcases a rudimentary Image Processing tool using the following SMART features: text and shape displays, touch area, switching between the three Camera Sensor modes of Motion Detection, Line Detection and Color Sensing with the goal of highlighting the mobile screen's zone where a user-chosen colored dowel can be found in real-time (*as much as possible*). The corresponding PLAY700 App project uses the following list of Text Items:

1 – Black

2 – Red

3 – Green

4 – Blue

5 – Scan

6 – Found

7 – Dowel of Interest

8 – Motion Detected at Red Circle

Fig. 1.46 describes the Initialization section of program "AD- ColorDowelAlert.tskx":

1) Lines 5 to 8 initialize three Parameters "ColorType" (Black, Red, Green or Blue), ScanMode (True/False) and ScanDone (True/False).

2) The robot then waits for the user to make a noise louder than 50 (Line 10).

3) Next, the mobile device is set to use the Back Camera, to set the screen in Portrait Mode, and clears the Text Display overlay (Lines 12-14).

4) Function "Menu" is called at Line 15. Details of Function "Menu" are shown in Fig. 1.48.

```
6   ColorType = 0
7   ScanMode = FALSE (0)
8   ScanDone = FALSE (0)
9   // User needs to make sure that robot is connected to mot
10  WAIT WHILE ( SMART: Noise (dB) < 50 )
11  // Getting remote device ready using SMART commands
12  SMART: Camera Selection = Back Camera (1)
13  SMART: Screen Rotation = Portrait Mode (1)
14  SMART: Text Display = 0
15  CALL Menu
```

Fig. 1.46 Initialization Part of Main Program in "AD-ColorDowelAlert.tskx".

```
17 ENDLESS LOOP
18 {
19     IF ( ColorType == 0 || ScanMode == FALSE (0) )
20     {
21         CALL CheckTouch
22     }
23     IF ( ScanMode == TRUE (1) )
24     {
25         SMART: Camera Sensor = Not Used (0)
26         CALL Delay2
27         ScanDone = FALSE (0)
28         CALL ScanDowel
29     }
30     IF ( ScanDone == TRUE (1) )
31     {
32         ColorType = 0
33         ScanMode = FALSE (0)
34         ScanDone = FALSE (0)
35         CALL Menu
36     }
37 }
```

Fig. 1.47 Main Endless Loop in "AD-ColorDowelAlert.tskx".

The Main Endless Loop (Fig. 1.47) is then entered and three parallel IF structures are executed for each loop iteration:

1) The first IF structure (Lines 19-22) checks to see if the 2 parameters "ColorType" and "ScanMode" are still at their initial values (i.e. "0" and "FALSE"), meaning that the user has not finished making his or her choice to scan for which Dowel Color (Black, Red, Green or Blue). If any of these 2 parameters is still at its initial value, then Function "CheckTouch" is called (described in Figs. 1.49 and 1.50). If "ColorType" is set to a proper value and "ScanMode" set to TRUE from inside Function "CheckTouch", then the second IF structure would be entered next.

2) The second IF structure (Lines 23-29) checks to see if "ScanMode" is set to "TRUE". If "ScanMode" is still set at its initial value "FALSE", the third IF structure is tested next. If (ScanMode == TRUE), the SMART Camera Sensor Mode is reset to zero (Line 25) and a delay of 250 ms is called (Line 26), essentially waiting for the mobile hardware/software to "catch up" with the CM-150 code at runtime. Next "ScanDone" is reset to FALSE (Line 27), i.e. getting ready to call Function "ScanDowel" in Line 28. Details for Function "ScanDowel" are shown in Figs. 51-53. When a successful detection for the dowel is done inside Function "ScanDowel", Parameter "ScanDone" is set to TRUE to activate the third IF structure.

3) The third IF structure (Lines 30-36) checks to see if "ScanDone" is set to "TRUE". If "ScanDone" is still set at its initial value "FALSE", the Endless Loop is just restarted next. If (ScanDone == TRUE),

then the 3 Parameters ColorType, ScanMode, and ScanDone are reset to their initial values (Lines 32-34). Next, Function Menu is called to restore the initial look of the Menu (Line 35), and the CM-150 program counter is sent back to the beginning of the Main Endless Loop in Fig. 1.47.

Fig. 1.48 shows the details of Function "Menu":

1) Lines 42 ad 43 clear the Text and Shapes Display overlays respectively.

2) Next the 5 Text Items (Black/Red/Green/Blue/Scan) are written in White Font at their assigned screen locations (Lines 44-48).

```
40  FUNCTION Menu
41  {
42      SMART: Text Display = 0
43      SMART: Shapes Display = 0
44      SMART: Text Display = [Position:(1,1)],[Item:1],[Size:40],[Color:White]
45      SMART: Text Display = [Position:(3,1)],[Item:2],[Size:40],[Color:White]
46      SMART: Text Display = [Position:(4,1)],[Item:3],[Size:40],[Color:White]
47      SMART: Text Display = [Position:(5,1)],[Item:4],[Size:40],[Color:White]
48      SMART: Text Display = [Position:(3,5)],[Item:5],[Size:40],[Color:White]
49  }
```

Fig. 1.48 Function "Menu" in "AD-ColorDowelAlert.tskx".

```
51 FUNCTION CheckTouch
52 {
53     Touch1 = SMART: Touch Area 1
54     IF ( Touch1 != 0 )
55     {
56         WAIT WHILE ( SMART: Touch Area 1 != 0 )
57         IF ( Touch1 == [Position:(1,1)] )
58         {
59             ColorType = Black Color (2)
60             SMART: Text Display = [Position:(1,1)],[Item:1],[Size:40],[Color:Red]
61             SMART: Text to Speech (TTS) = Text Item 1
62             WAIT WHILE ( SMART: Text to Speech (TTS) != 0 )
63         }
64         ELSE IF ( Touch1 == [Position:(3,1)] )
65         {
66             ColorType = Red Color (3)
67             SMART: Text Display = [Position:(3,1)],[Item:2],[Size:40],[Color:Red]
68             SMART: Text to Speech (TTS) = Text Item 2
69             WAIT WHILE ( SMART: Text to Speech (TTS) != 0 )
70         }
```

Fig. 1.49 Part 1 of Function "CheckTouch" in "AD-ColorDowelAlert.tskx".

Figs. 1.49 and 1.50 show the details of Function "CheckTouch":

1) Line 53 saves the value of the detected Touch Area 1 into Parameter Touch1. If the user had pressed somewhere on the mobile screen, "Touch1" would have a value between 1 and 25, otherwise "Touch1" has a value of zero.

2) Line 54 checks to see if the user had pressed on the mobile screen, i.e. when the expression (Touch1 != 0) evaluates to TRUE, then Line 56 is used to wait for the user to lift his or her finger of the mobile screen.

3) Next, an IF-ELSE-IF structure (Lines 57-91) is used to figure out the 5 possible cases where the user has pressed on the mobile screen:

- If Position (1,1) is pressed (Line 57), this means that the user wants to track a Black dowel, so "ColorType" is set to Black Color (Line 59) and "Black" (Text Item 1) is rewritten in Red Font (Line 67). Furthermore Lines 68-69 verbalize the Text Item 1 (i.e. "Black").

- If Position (3,1) is pressed (Line 64), this means that the user wants to track a Red dowel, so "ColorType" is set to Red Color (Line 66) and "Red" (Text Item 2) is rewritten in Red Font (Line 67). Lines 68-69 additionally verbalize the Text Item 2 (i.e. "Red").

```
71 ELSE IF ( Touch1 == [Position:(4,1)] )
72 {
73     ColorType = Green Color (4)
74     SMART: Text Display = [Position:(4,1)],[Item:3],[Size:40],[Color:Red]
75     SMART: Text to Speech (TTS) = Text Item 3
76     WAIT WHILE ( SMART: Text to Speech (TTS) != 0 )
77 }
78 ELSE IF ( Touch1 == [Position:(5,1)] )
79 {
80     ColorType = Blue Color (5)
81     SMART: Text Display = [Position:(5,1)],[Item:4],[Size:40],[Color:Red]
82     SMART: Text to Speech (TTS) = Text Item 4
83     WAIT WHILE ( SMART: Text to Speech (TTS) != 0 )
84 }
85 ELSE IF ( Touch1 == [Position:(3,5)] )
86 {
87     SMART: Text Display = [Position:(3,5)],[Item:5],[Size:40],[Color:Red]
88     SMART: Text to Speech (TTS) = Text Item 5
89     WAIT WHILE ( SMART: Text to Speech (TTS) != 0 )
90     ScanMode = TRUE (1)
91 }
```

Fig. 1.50 Part 2 of Function "CheckTouch" in "AD-ColorDowelAlert.tskx".

- If Position (4,1) is pressed (Line 71 in Fig. 1.50), this means that the user wants to track a Green dowel, so "ColorType" is set to Green Color (Line 73) and "Green" (Text Item 3) is rewritten in Red Font (Line 74). Furthermore Lines 75-76 verbalize the Text Item 3 "Green".
- If Position (5,1) is pressed (Line 78), this means that the user wants to track a Blue dowel, so "ColorType" is set to Blue Color (Line 80) and Text Item 4 "Blue" is rewritten in Red Font (Line 81). Lines 82-83 verbalize Text Item 4 (i.e. "Blue").
- If Position (3,5) is pressed (Line 85), this means that the user wants to start the Scanning Process of a given dowel, so Text Item 5 "Scan" is rewritten in Red Font (Line 87) and Lines 88-89 verbalize Text Item 5 (i.e. "Scan"), and Parameter "ScanMode" is set to TRUE (getting ready for the second IF structure in Fig. 1.47).

Figs. 1.51, 1.52 and 1.53 show the details of Function "ScanDowel" which has multiple layers of "sequential logic":

1) First, the Motion Detection Mode is set at line 98, followed by a delay of 250 ms (Line 99) to wait for the mobile device hardware/software to catch up with the logic flow inside the TASK program

(running on the CM-150). The user can comment out Line 99 and sees the effect of this small delay on the overall performance of this program.

2) Next, the current value from the Motion Detection Area Function is saved into Parameter "Detected_Zone" (Line 101). If no Motion is perceived by the Camera, "Detected_Zone" has a value of zero, otherwise "Detected_Zone" would have a numerical value between 1 and 25.

```
95  FUNCTION ScanDowel
96  {
97     // Set Motion Detection mode
98     SMART: Camera Sensor = Motion Detection Mode (3)
99     CALL Delay2
100    // Motion detection in Detected_Zone 1 through 25
101    Detected_Zone = SMART: Motion Detection Area
102    IF ( Detected_Zone != 0 )
103    {
104       Dowel_Zone = 0
105       SMART: Camera Sensor = Not Used (0)
106       CALL Delay2
107       // Switching to Line Detection mode
108       SMART: Camera Sensor = Line Detection Mode (4)
109       SMART: Tracking Color (Line-Tracer) = ColorType
110       CALL Delay2
111       Dowel_Zone = SMART: Line Detection Area
112       IF ( Dowel_Zone != 0 )
113       {
114          SMART: Shapes Display = [Position:Dowel_Zone],[Item:3],[Size:60],[Color:Green]
115          SMART: Text Display = [Position:(3,3)],[Item:7],[Size:40],[Color:Yellow]
116          SMART: Text Display = [Position:(3,4)],[Item:6],[Size:40],[Color:Yellow]
117          SMART: Text to Speech (TTS) = Text Item 7
118          WAIT WHILE ( SMART: Text to Speech (TTS) != 0 )
119          SMART: Text to Speech (TTS) = Text Item 6
120          WAIT WHILE ( SMART: Text to Speech (TTS) != 0 )
121          CALL Delay1
122          SMART: Text Display = [Position:(3,3)],[Item:0],[Size:40],[Color:Black]
123          SMART: Text Display = [Position:(3,4)],[Item:0],[Size:40],[Color:Black]
124          SMART: Shapes Display = [Position:Dowel_Zone],[Item:0],[Size:60],[Color:Unknown]
125          ScanDone = TRUE (1)
126       }
```

Fig. 1.51 Part 1 of Function "ScanDowel" in "AD-ColorDowelAlert.tskx".

3) Thus if (Detected_Zone != 0) evaluated to TRUE (Line 102) (i.e. Motion is perceived), then we need to investigate further if the Perceived Zone contains pixels that are mostly of the Dowel Color that the user has specified before. Therefore, we need to switch to Color Line Detection Mode next (Lines 104-111).

4) Line 104 resets Parameter "Dowel_Zone" to zero.

5) At this point, the Camera Sensor Mode was already set as Motion Detection Mode (Line 98), so we need to reset it to Mode 0 (Line 105) for 250 ms (Line 106) – to be on the safe side – before setting it to the new Line Detection Mode (Line 108). Parameter "ColorType" previously set by the user in Function "CheckTouch" is now used to set the Function SMART Tracking Line Color (Line Tracer) in Line 109. Another 250 ms delay is again used to let the mobile device keep pace with the TASK program (Line 110).

6) We now can retrieve numerical results from the SMART Line Detection Area process and save them into Parameter Dowel_Zone which can be from 1 to 5 (if pixels of "ColorType" values are found in the top fifth of the mobile screen, i.e. Zone 1 to Zone 5), or zero (if no pixels of "ColorType" values are found in these 5 zones) – this is the job of Line 111.

7) Next, the IF structure (Lines 112 to 126) is used to perform several actions if "Dowel_Zone" turns out to be non-zero:

- A Green Triangle (Shape 3) is drawn at the screen Zone defined by the value of "Dowel_Zone" (Line 114).

- Text Item 7 "Dowel of Interest" and Text Item 6 "Found" are displayed in Yellow Font (Lines 115 and 116) and additionally verbalized (Lines 117 to 120). A 1 second delay is used on Line 121 to let the user hear and read the prompts on the mobile device screen. Lines 122-124 are used to clear the Text and Shapes Display Overlays.

- Parameter "Scan_Done" is then set to TRUE as a return value of the Function "Scan_Dowel" to let the Main Endless Loop know about the "successful" result of the Dowel Scanning process.

8) If "Dowel_Zone" turns out to be ZERO instead, i.e. no pixels of "ColorType" values are found within Zones 1 through 5, then the ELSE structure starting from Line 127 is activated next (see Fig. 1.52):

- At this point the only PLAY700 tool left to use is the Color Detection tool which monitors only the pixels within Zone 13, i.e. Position (3,3). Thus, first Parameter "ColorFound" is set to zero (Line 129).

- Then SMART Camera Sensor Mode is set to zero and a 250 ms is used to better synchronize between the mobile device and the CM-150 (Lines 130-131).

- Line 133 sets the Camera Sensor to Color Detection Mode with another 250 ms delay (Line 134).

- With Line 135, the SMART Detected Color Function is used to return the average color value of all the pixels in Zone 13 and save it in Parameter "ColorFound".

- Next, the IF structure defined between Lines 136 and 150 is used to perform several actions if "ColorFound" turns out to match with Parameter "ColorType" (which was chosen by the user way back in Function "CheckTouch", see Figs. 1.49 and 1.50) – this is the job of Line 136. If the match is TRUE, then a Green Square is drawn at Position (3,3) (Line 138), next Text Items 7 and 6 are displayed in Yellow (Lines 139-140) and then verbalized (Lines 141-144).

```
127       ELSE
128       {
129          ColorFound  =  0
130          SMART: Camera Sensor  =  Not Used (0)
131          CALL  Delay2
132          //  Switching to Color Detection mode using Zone (3,3)
133          SMART: Camera Sensor  =  Color Detection Mode (2)
134          CALL  Delay2
135          ColorFound  =  SMART: Detected Color
136          IF  (  ColorFound  ==  ColorType  )
137          {
138             SMART: Shapes Display  =  [Position:(3,3)],[Item:2],[Size:60],[Color:Green]
139             SMART: Text Display  =  [Position:(2,4)],[Item:7],[Size:40],[Color:Yellow]
140             SMART: Text Display  =  [Position:(4,4)],[Item:6],[Size:40],[Color:Yellow]
141             SMART: Text to Speech (TTS)  =  Text Item 7
142             WAIT WHILE  (  SMART: Text to Speech (TTS)  !=  0  )
143             SMART: Text to Speech (TTS)  =  Text Item 6
144             WAIT WHILE  (  SMART: Text to Speech (TTS)  !=  0  )
145             CALL  Delay1
146             SMART: Text Display  =  [Position:(2,4)],[Item:0],[Size:40],[Color:Black]
147             SMART: Text Display  =  [Position:(4,4)],[Item:0],[Size:40],[Color:Black]
148             SMART: Shapes Display  =  [Position:(3,3)],[Item:0],[Size:60],[Color:Unknown]
149             ScanDone  =  TRUE (1)
150          }
```

Fig. 1.52 Part 2 of Function "ScanDowel" in "AD-ColorDowelAlert.tskx".

- Line 145 provides a 1 second delay to allow the user to hear and read the screen prompts. Then the Text and Shapes Display overlays are cleared by Lines 146-148.
- Similarly, Parameter "Scan_Done" is then set to TRUE as a return value of the Function "Scan_Dowel" to let the Main Endless Loop know about the "successful" result of the Dowel Scanning process (Line 149).

The last part for Function "ScanDowel" is shown in Fig. 1.53 which is for the case where we know that Motion got detected, i.e. Detected_Zone is non-zero, but it is not Zone 1 through 5, nor is it Zone 13:

```
151         ELSE
152         {
153             SMART: Shapes Display = [Position:Detected_Zone],[Item:1],[Size:60],[Color:Red]
154             SMART: Text Display = [Position:(3,3)],[Item:8],[Size:30],[Color:Yellow]
155             SMART: Text to Speech (TTS) = Text Item 8
156             WAIT WHILE ( SMART: Text to Speech (TTS) != 0 )
157             CALL Delay1
158             SMART: Text Display = [Position:(3,3)],[Item:0],[Size:30],[Color:Black]
159             SMART: Shapes Display = [Position:Detected_Zone],[Item:0],[Size:60],[Color:Unknown]
160             ScanDone = TRUE (1)
161         }
162     }
```

Fig. 1.53 Part 3 of Function "ScanDowel" in "AD-ColorDowelAlert.tskx".

1) Then a Red Circle is drawn at the Position defined by Parameter "Detected_Zone", determined way back in Line 101 of Fig. 1.51.

2) Text Item 8 "Motion Detected at Red Circle" is displayed and verbalized (Lines 154-156).

3) A 1 second delay is used to allow the user to hear and read the screen prompts (Lines 158-159).

4) Finally, Parameter "ScanDone" is set to TRUE and the TASK program counter returns to Line 29 of the Main Endless Loop in Fig. 1.47, and this TASK program is ready for the user to choose a new color to scan for.

Video 1.11 shows the runtime performance of the program "AD-ColorDowelAlert.tskx".

The PLAY700 App has many other tools such as Gesture, SMS Receiver, Email Sender (to a Google Mail account only) which are outside the scope this book, but example programs illustrating their uses are included with this book, please visit the web page www.cntrobotics.com/advanced_dream for instructions about how to get them. The original codes are in Korean and they can be obtained here (http://support.robotis.com/ko/software/roboplus/roboplus_task/programming/parameter/remoteterminal/smart_app_parameter.htm#MainSpec44), the English speaking reader needs to use Chrome so that the Korean contents can be translated into English.

1.9 Dual CM-150 Programming via Bluetooth

Eventually, the reader would find the two GPIO Ports (3 and 4) on the CM-150 as quite limiting and one may wonder if a robot with multiple CM-150 controllers can be built and most importantly programmed to work together collaboratively. And the answer is Yes, but for now between 2 CM-150s only, when using Bluetooth (BT-210 or BT-410 – in theory) because so far ROBOTIS has only released Bluetooth firmware that can only deal with a 1 to 1 pairing scheme, Master-Slave to be exact, and with a 1 to N version to be released sometime in the future (http://emanual.robotis.com/docs/en/parts/communication/bt-210/, http://emanual.robotis.com/docs/en/parts/communication/bt-410/).

The author would add that, presently when using TASK, ones can coordinate **multiple** ROBOTIS CM-XXX controllers to work together ONLY using ZigBee communications hardware (http://emanual.robotis.com/docs/en/parts/communication/zig-110/, http://www.robotis.us/zig-110a-set/) – unfortunately these ZigBee hardware are discontinued so they are hard to obtain nowadays. Thai (2017) devoted Chapter 7 of that book to the topic of "Communication Programming with REMOCON Packets" and the concept of "Message-Shaping" as it applied to ZigBee communications hardware and protocols (i.e. 1 to 1 and broadcast modes). This section 1.9 applies this "Message-Shaping" concept to a Bluetooth Master-Slave configuration between two CM-150 controllers.

Currently, ROBOTIS offers a set of Master/Slave BT-210s (http://www.robotis.us/bt-210-set/) and another set of Master/Slave BT-410s (http://www.robotis.us/bt-410-set/), however the author has found that only the **BT-210 Master/Slave set** would work properly for TASK codes between 2 CM-150s as developed in this Section 1.9. If the reader happens to have 2 BT-210s available and both are in Slave mode, then the user can follow the procedure described at this web link (http://emanual.robotis.com/docs/en/parts/communication/bt-210/) to convert them into a Master/Slave set. Unfortunately, the reader will also need to use an OpenCM-9.04-C controller (http://www.robotis.us/opencm9-04-c-with-onboard-xl-type-connectors/) in order to perform the necessary steps to set one BT-210 into a Master mode and also to pair it with the other BT-210 in Slave mode.

In Section 3.7 of Thai (2018), the author had provided some programming techniques to handle the 10 bits of a typical REMOCON message transmitted from a physical or virtual remote controller RC-100A/100B (as it has only 10 buttons). However, the "full" REMOCON message has 16 bits that the user can program to accommodate more advanced communications needs. Please visit this web link (http://emanual.robotis.com/docs/en/parts/communication/rc-100/#communication-packet) for complete details of the RC-100 communication packet, also referred to as REMOCON packets.

The reader may remember that most ROBOTIS sensors have digitized values between 0 and 1023, and this range of values can be dealt with using only the 10 low bits of the 16 bits REMOCON message. This means that the user can use the highest 6 bits for other purposes: this is the approach used in the following project which has a Master CM-150 robot sending its NIR sensors values via Bluetooth which are to be used by a separate Slave CM-150, which can also use its own NIR sensors for its programmed maneuvers. The relevant TASK codes are named "AD-RemoteSensingFollower_M.tskx" for the Master robot and "AD-RemoteSensingFollower_S.tskx" for the Slave robot.

For this project, the 16-bit REMOCON message is organized as follows:

- The lower 12 bits (i.e. Bit 0 through Bit 11) will be used for the actual Sensor Value (only 10 bits are actually needed).
- The higher 4 bits (i.e. Bit 12 through Bit 15) will be used to denote the Sensor ID. Please note that with 4 bits, we can have up to $2^4 - 1 = 15$ distinct sensors which should be adequate between 2 CM-150s. Of course, this technique when applied to more advanced controllers like the CM-200 or CM-530 would require more Sensor IDs, then the reader can use the 10 lowest bits for Sensor Data and the highest 6 bits for Sensor IDs (i.e. $2^6 - 1 = 63$ distinct sensors or actuators).

When the reader browses through the two sample TASK codes for Master and Slave, the reader should note that both programs share a common algorithm with the usual Main Endless loop which has 4 sections:

1. The first section shows how a current sensor reading of the local controller is shaped into a message that also contains that sensor's ID and how this combined message is sent to the remote controller.

2. The second section describes how the local controller receives a remote message and how it decodes a given remote message to recover both the remote Sensor ID and Sensor Value embedded in the message received.
3. The third section illustrates how the local controller processes the remote sensor information together with its local sensor data to perform programmed behaviors for the local robot.
4. The fourth section is simply a tuning loop to get the rate of sending and receiving messages not too high to overwhelm the Bluetooth communication channel.

Fig. 1.54 shows how the Main Loop for the Master bot starts (i.e. Part 1 of Master/Slave algorithm):

- Line 17 shows how the Master controller saves its current IR_Left Sensor Value into Parameter Master_Sensor_1.
- Line 19 shows how to embed this Sensor ID, which happens to be "1", starting from Bit 12 of Parameter "Message1" (please note that customarily the right-most bit is labeled Bit 0 and the neighboring bit to its left is labeled Bit 1, and so forth). The RHS of the assignment command on Line 19 shows that ID "1" is multiplied by the binary number "1 0000 0000 0000" (or 4096 in decimal) in order to affect essentially a left-shift of 12 bits for the ID value of "1". The TASK tool does not support the usual left-shift "<<" operator thus we have to fall back upon the basic Multiplication "*" operator.

```
13    // Master's L-C-R IR sensors will be Remote Sensor 1, 2, 3 to the Slave, and vice-versa
14    ENDLESS LOOP
15    {
16        // Collect current Master's sensors data and send them to Slave via Remocon packets
17        Master_Sensor_1 = IR Left
18        // Shift Sensor ID = 1 to Bit 12 (counting from Bit 0)
19        Message1 = 0000 0000 0000 0000 0001 0000 0000 0000 * 1
20        // Combine Master_Sensor_1 into Message1 from Bit 0 to Bit 11 (1023 only requires 10 bits)
21        Message1 = Message1 + Master_Sensor_1
22        // And send it away to Slave with TXD command
23        Remocon TXD = Message1
24        // Repeat procedure for Master Sensor 2
25        Master_Sensor_2 = IR Center
26        Message2 = 0000 0000 0000 0000 0001 0000 0000 0000 * 2
27        Message2 = Message2 + Master_Sensor_2
28        Remocon TXD = Message2
29        // finally, repeat procedure for Master Sensor 3
30        Master_Sensor_3 = IR Right
31        Message3 = 0000 0000 0000 0000 0001 0000 0000 0000 * 3
32        Message3 = Message3 + Master_Sensor_3
33        Remocon TXD = Message3
```

Fig. 1.54 Part 1 of Main Loop for **Master Bot** in "**AD-RemoteSensingFollower_M.tskx**".

- Line 21 then shows how an Addition operation combines the Sensor ID in the highest 4 bits of Message1 and the Sensor Value in the lowest 12 bits of Message1.
- Line 23 then sends this combined message to the Slave controller via the "Remocon TXD" command.
- Lines 25 through 28 repeat the above procedure for Sensor ID "2" and its corresponding IR_Center Sensor Value.
- And Lines 30 through 33 then complete the sending of Message3 for the IR_Right Sensor.
- In short, during each iteration of its Main Loop, the Master controller sends out 3 messages to the Slave controller.

Next, Fig. 1.55 shows how the Slave controller decodes these 3 messages and save the information received in the proper parameters (i.e. Part 2 of general Master/Slave algorithm):

```
35  // Receive sensors data from Master and apply them to Slave's maneuvering scheme
36  IF ( Remocon Data Received == TRUE (1) )
37  {
38      MessageReceived = Remocon RXD
39      // Filtering out Bits 12-15 to find SensorID part of MessageReceived
40      SensorID = MessageReceived & 0000 0000 0000 0000 1111 0000 0000 0000
41      // Map SensorID back down to Bits 0-3, i.e. shift right 12 bits
42      SensorID = SensorID / 0000 0000 0000 0000 0001 0000 0000 0000
43      // Filtering out the lowest 12 bits (Bits 0-11) to get the actual Sensor Value part of the message received
44      SensorValue = MessageReceived & 0000 0000 0000 0000 0000 1111 1111 1111
45      IF ( SensorID == 1 )
46      {
47          Remote_Sensor_1 = SensorValue
48      }
49      ELSE IF ( SensorID == 2 )
50      {
51          Remote_Sensor_2 = SensorValue
52      }
53      ELSE IF ( SensorID == 3 )
54      {
55          Remote_Sensor_3 = SensorValue
56      }
57  }
```

Fig. 1.55 Part 2 of Main Loop for **Slave Bot** in "**AD-RemoteSensingFollower_S.tskx**".

- When a CM-150 receives a new Remocon packet, its flag "Remocon Data Received" is set to TRUE, thus the use of the IF structure started on Line 36 in Fig. 1.55.
- Once a new Remocon packet is received, the message (previously sent by the Master CM-150) is saved into Parameter "MessageReceived" by the Slave CM-150 (Line 38).
- Line 40 extracts out the Sensor ID component by using the bit-wise AND operator "&" between "MessageReceived" and the binary mask "1111 0000 0000 0000" – i.e. only the highest 4 bits of the 16-bit message are extracted – and saves this result in Parameter "SensorID". However,

this Sensor ID value is still located in the upper 4 bits of Parameter "SensorID", thus this information not yet usable.

- Next, Line 42 uses a "Division by 1 0000 0000" operation upon Parameter "SensorID" to shift this information "right" 12 bits so that the Sensor ID information is finally a correct numerical value to use (i.e. "SensorID" value would be either 1, 2 or 3).
- Lastly, Line 44 extracts out the information found in the lowest 12 bits of "MessageReceived" and saves it into Parameter "SensorValue" by using another bit-wise AND operation with the constant "0000 1111 1111 1111".
- Lines 45-56 form an IF-ELSE-IF structure designed to save "SensorValue" into the proper parameters "Remote_Sensor_1", "Remote_Sensor_2" or "Remote_Sensor_3" depending on the actual value received for "SensorID" (1, 2 or 3).

The Slave controller is now ready to use the information from the remote and local sensors to affect its actuators as shown in Fig. 1.56 (i.e. Part 3 of general Master/Slave algorithm):

```
// Applying Slave's Sensor Data to Slave's maneuvers
IF ( Slave_Sensor_2 >= 50 )
{
    CALL Backward
}
ELSE
{
    // Applying Master's Sensor Data to Slave's maneuvers - Remote Sensor 2 is not used
    IF ( Remote_Sensor_1 >= 5 && Remote_Sensor_3 < 2 )
    {
        CALL Left
    }
    ELSE IF ( Remote_Sensor_3 >= 5 && Remote_Sensor_1 < 2 )
    {
        CALL Right
    }
    ELSE IF ( Remote_Sensor_1 >= 40 && Remote_Sensor_3 >= 40 )
    {
        CALL Forward
    }
    ELSE
    {
        CALL Stop
    }
}
```

Fig. 1.56 Part 3 of Main Loop for **Slave Bot** in "**AD-RemoteSensingFollower_S.tskx**".

- Overall, an IF-ELSE-IF structure (Lines 66-81) is embedded in an outer IF-ELSE structure to reflect the higher priority of the Local Sensor "Slave_Sensor_2" (i.e. Slave's IR_Center) over the Remote_Sensors (1/2/3).
- Lines 59-62 show that if Slave_Sensor_2 (built-in IR_Center) is greater or equal than 50, the Slave robot should back away from the very real obstacle in front of it. Otherwise (Lines 63-82), the Slave should act upon the information received from the Remote Sensors: in this situation, ones can think as if the Slave robot reacts to a "virtual" obstacle while the "real" obstacle is parading in front of the Master robot!
- Lines 66-81 is the standard Follower algorithm previously detailed in Thai (2018) whereas:
  - If there is a "real" obstacle on the "Left of the Master Bot", then the "Slave Bot" should turn Left into it, i.e. to track a "virtual" obstacle! (Lines 66-69).
  - Else If there is a "real" obstacle on the "Right of the Master Bot", then the "Slave Bot" should turn Right into it to track it "virtually"! (Lines 70-73).
  - Else If there is a "real" obstacle in "Front of the Master Bot", the "Slave Bot" should go Forward to follow the virtual obstacle (Lines 74-77).
  - In all other cases, the Slave Bot should stay still (Lines 78-81).

Fig. 1.57 represents a small counter-based loop designed to create a user-tunable time delay so that the rate of sending/receiving packets is not too fast to overwhelm the Bluetooth communication channel (Part 4 of general Master/Slave algorithm).

```
77      // Have a short delay so as NOT to send/receive messages too quickly one after another - Adjust for your own needs
78      Delay_Counter = 50
79      LOOP WHILE ( Delay_Counter != 0 )
80      {
81          Delay_Counter = Delay_Counter - 1
82      }
83  }
```

Fig. 1.57 Part 4 of Main Loop for **Master Bot** in "**AD-RemoteSensingFollower_M.tskx**".

When the reader browses through the complete sample code "AD-RemoteSensingFollower_M.tskx", he/she would realize that there is a code section written for the Master Controller to receive and decode the remote messages from the Slave Controller which is identical to the one shown in Fig. 1.55. Unfortunately, this code section, currently, would not be activated for the Master Bot at all, because the current BT-210 firmware is ONLY allowing the Master to send packets and not to receive them (or may be the combination of the current BT-210 and CM-150 firmware?). Hopefully, in the future, ROBOTIS would release a new BT-210 firmware that would allow true bi-directional flow of packets.

However, the author had found an interesting side effect when the code sections for checking the "Remocon Data Received" flag is removed from both Master's and Slave's program (Lines 36-37 and 57 in Fig. 1.55), then the Master Bot can now receive Remote Sensor messages from the Slave, but now the Slave cannot receive any Remote Sensor messages from the Master! For more details, please see enclosed sample codes "AD-RemoteSensingFollower_M_R.tskx" and "AD-RemoteSensingFollower_S_R.tskx". Thus, it seems that the TXD packets are physically sent out by both robots, but current firmware is not allowing a true bi-directional flow of information between 2 CM-150s via BT-210 YET!

Video 1.12 illustrates the run-time performances of both versions of the sample codes for Master and Slave robots.

Currently, the reader interested in true multiple-robots control would need to use the Edbot tool to be described in the coming Chapter 2.

1.10 Use of Sample Codes on Mobile Devices

All sample TASK codes for this book can also be transferred to an Android or iOS device so that they can be used by the mobile-equivalent of the TASK tool, called R+m.TASK2, available at (https://play.google.com/store/apps/details?id=com.robotis.task2&hl=en) for the Android version or at (https://itunes.apple.com/us/app/r-m-task2-robotis/id1031166481?mt=8) for the iOS version. To transfer the sample TSK codes to an Android device, the user only needs to use Windows Explorer on the PC side and then drags and drops them into the appropriate folder on the mobile device (see Fig. 1.58). For an iOS device, the user will need to use iTunes to transfer the sample codes from the PC to the user personal iOS device, via File Sharing (see Fig. 1.59).

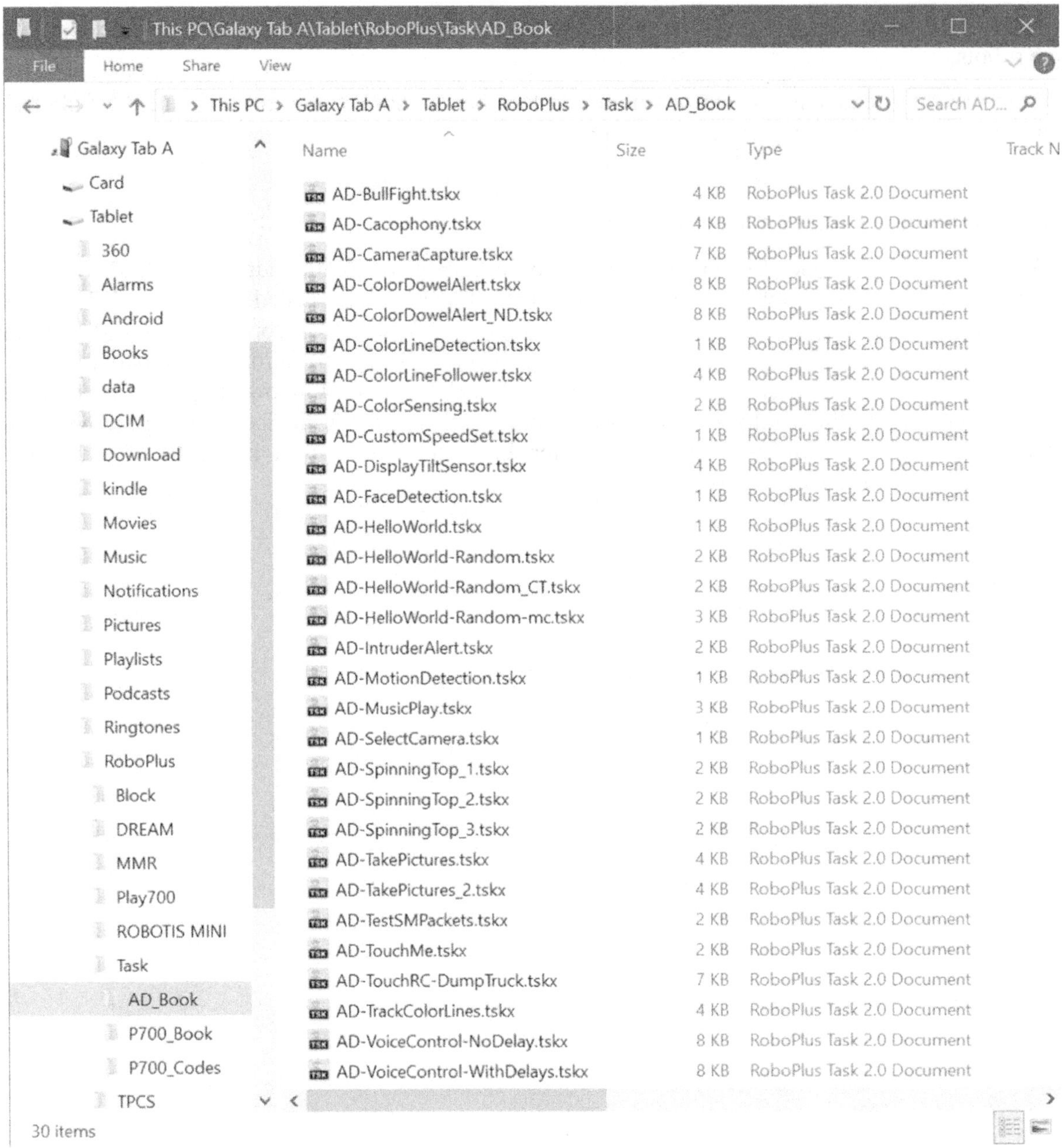

Fig. 1.58 Transferring sample codes to an Android device via Windows File Explorer.

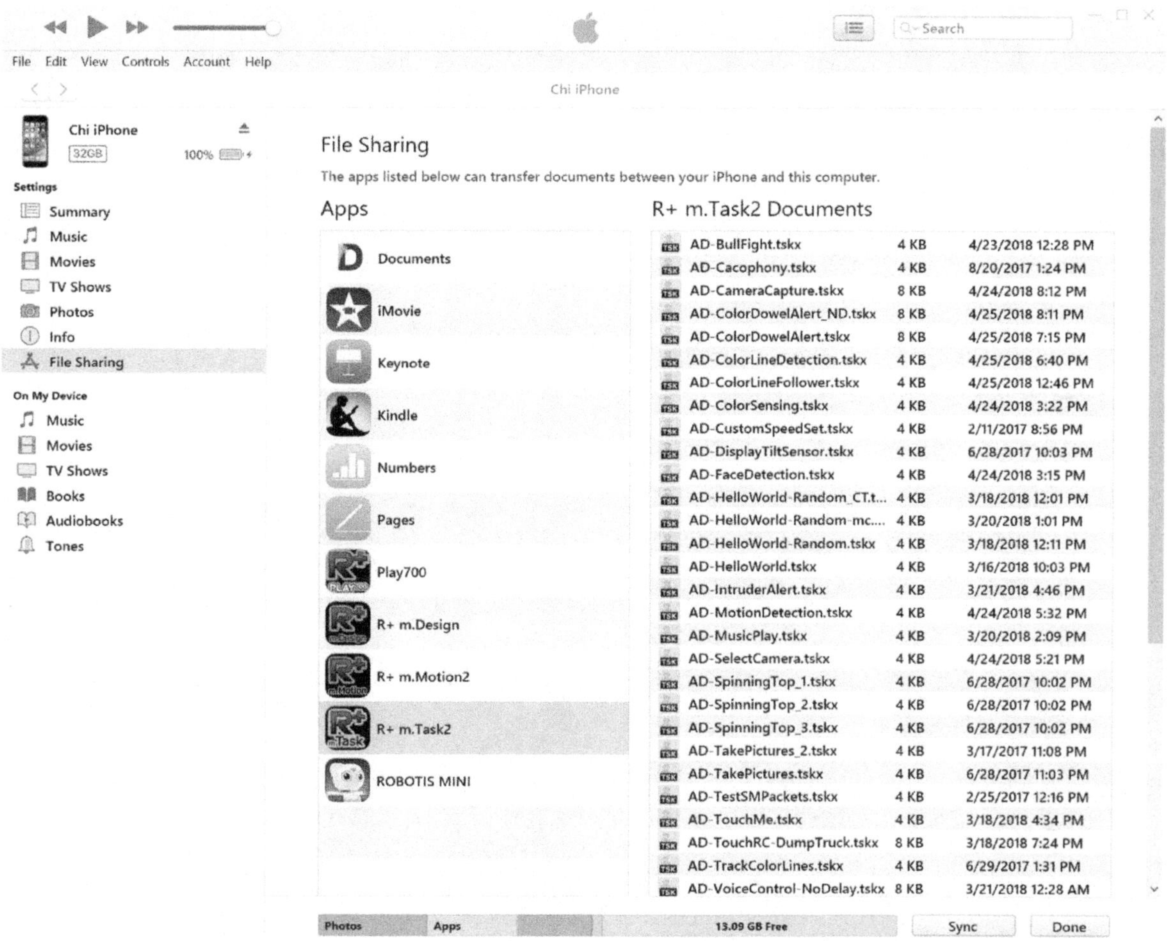

Fig. 1.59 Transferring sample codes to an iOS device via iTunes File Sharing.

1.11 References

Thai CN (2017) Exploring Robotics with ROBOTIS Systems (2nd Edition). Springer-Nature Publishers, Cham Switzerland.

Thai CN (2018) Learning Robotics with ROBOTIS DREAM System. CNT Robotics LLC Publisher. Printed by Amazon Create Space.

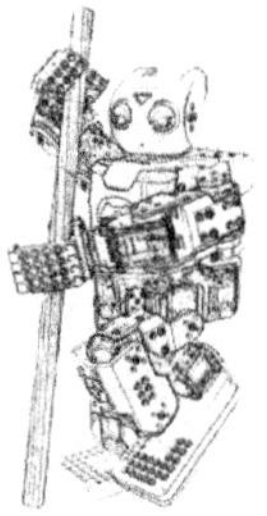

Chapter 2: Using Edbot® Dream Pro

Circa 2015, the Edbot software tool was first created as a server/client tool to be used within an actual classroom environment, along with teacher supervision of student access to individual ROBOTIS MINI robots (http://www.robotis.us/robotis-mini/), programmed with the MIT Scratch 2 Offline IDE (http://ed.bot/edbot/). In 2018, the Edbot suite was expanded with the release of the Edbot Dream tool (http://ed.bot/edbotdream/) which was designed to work with the DREAM II School Set (http://www.robotis.us/dream/).

In January 2019, SCRATCH 3 (https://en.scratch-wiki.info/wiki/Scratch_3.0) was released with an online interface and a desktop interface (https://scratch.mit.edu/download). In March 2019, Edbot V.5 was released allowing the additional usage of SCRATCH 3 via a web interface (http://scratch.ed.bot/) – a desktop version may be released in the future. Unfortunately, this web interface yields a slower communications speed with the DREAM robot and makes autonomous behaviors harder to achieve, thus this book concentrates on the usage of SCRATCH 2 Offline Editor instead. Please see Section 2.5 for more discussions regarding the use of SCRATCH 3 with the DREAM II robot.

Furthermore, the Edbot Dream Pro tool (V.5 and above) allows **multiple robot** programming via three software interfaces: Scratch, Python and JavaScript (http://support.ed.bot/edbot-dream-index.html), and with more languages such as Node.js, Java and C# and other platforms such as Chromebooks being planned.

In this book, an individual user is assumed, i.e. the user's PC is used as the local host (Port 8080) and the Edbot Dream Pro tool is used in its server mode. Please view this YouTube video for an overview of the capabilities of the Edbot Dream Pro tool (https://www.youtube.com/watch?v=pOk5szsngM0).

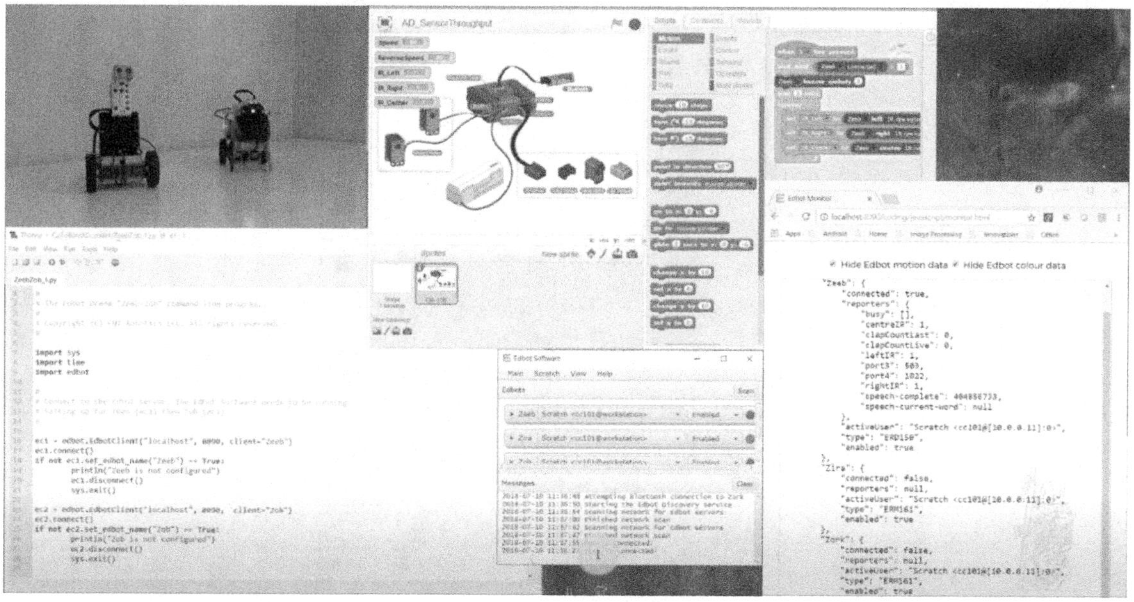

Fig. 2.1 Multi-software and multi-robot application of the Edbot Dream Pro tool.

Furthermore, the Edbot software environment allows the mixing of MINI and DREAM robots in a single instance of Scratch (https://www.youtube.com/watch?v=yUpx42eNR2I).

For this Chapter, the author assumes that the reader is already familiar with MIT SCRATCH 2 software or has read Chapter 4 of the other DREAM book by the author (Thai, 2018). If not, the user is recommended to first read up on such works as Ford (2014), Warner (2015) or Vlieg (2016) to have a thorough understanding of the SCRATCH 2 language. In this Chapter, only selected SCRATCH 2 features, most applicable to robotics and unique to the workings of the Edbot Dream Pro software, would be presented. However, Section 2.5 will present some "appropriate" SCRATCH 3 applications with the DREAM II robot.

2.1 Edbot's Hardware and Software Installation Requirements

The user will need to purchase the appropriate Edbot Dream kits and/or software product keys from Robots in School Ltd. (https://shop.ed.bot/collections/products). If the user has already purchased the DREAM II School Set elsewhere, the user can just buy the Edbot Pro product key from Robots in School Ltd. (https://shop.ed.bot/collections/products/products/edbot-software) or from ROBOTIS USA (http://www.robotis.us/edbot-software-product-key/). **In this book, the author assumes that the user is using the Pro version of the Edbot Dream tool (Version 5.0.6.1280 or higher).**

Hardware wise, the Edbot Dream software only works with either the ROBOTIS BT-210 module (http://www.robotis.us/bt-210/), or the BlueTooth module designed by Robots in School Ltd. (https://shop.ed.bot/collections/products/products/bluetooth-module-for-robotis-dream).

Please view Video 2.0 for the typical steps needed to install the Edbot Dream Pro software on the user's PC. **Importantly, each software product key will be bound permanently to the specific BT-210 or the Edbot BlueTooth modules used, thus the user will need to plan out carefully his or her "robots" organization if the user wants to use the Edbot software with multiple robots.** Fig. 2.2 shows the author's Edbot set up for 4 robots: Zeeb and Zob are DREAM robots, while Zira and Zork are MINI robots.

Fig. 2.2 Author's Edbot Setup for his 4 robots.

Lastly, the user needs to update the firmware of the DREAM controller CM-150 to Version 33 or above. Please refer to this YouTube video as a general procedure (https://www.youtube.com/watch?v=LX3WuWJQosw) using the MANAGER tool from ROBOTIS.

2.2 SCRATCH 2 Project File Conversion

The author used a DREAM robot named "Zeeb" to create the SCRATCH 2 sample codes presented in this Chapter (see Fig 2.3), but the reader may use a different name for his or her robot, and thus will need to do a conversion of the SCRATCH SB2 files provided by the author.

Fig. 2.3 Front view of author's DREAM robot named "Zeeb".

Let's say that the user's robot name is "Spot" and that the user has put the author's sample codes in a folder named "Author" and has prepared a blank folder named "User" to receive the converted SB2 files whereas the "Zeeb" name is converted to "Spot". Next, let's assume that the user has installed the EdBot software properly and has successfully registered his or her DREAM robot to the name "Spot" via a specific BT module. Thus, when the user runs the EdBot software, he or she would see a starting screen (Edbot V.4) as shown in Fig. 2.4, when the item "Scratch" has been chosen from the Top Menu bar.

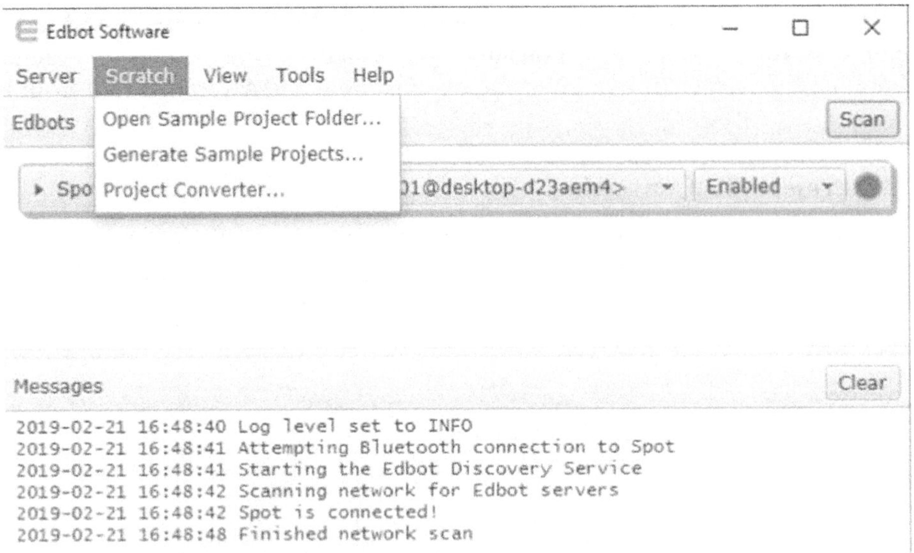

Fig. 2.4 Starting Screen of EdBot (V.4) software with item "Scratch" chosen on Top Menu bar.

Next, the user should choose the item "Project Converter" and a new window pops up (Fig. 2.5) where the user can browse to select a given Author's SB2 file to be converted as "Source File". If the button "Different Target File" is checked, then the user can specify a different file location and name for the "Target File". If not, the newly converted Target File will overwrite the original Source File.

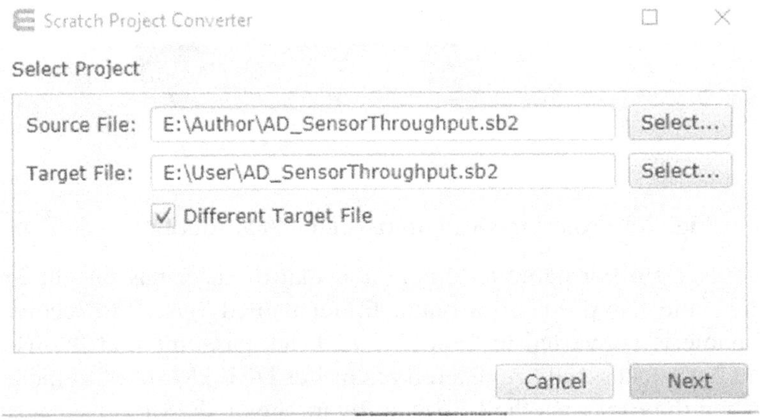

Fig. 2.5 Choosing "Source File" and "Target File" in "Project Converter".

When the user clicks on "Next" as shown in Fig. 2.5, the "Set Edbot Names" window pops up (see Fig. 2.6) where the user can verify that the Project Name "Zeeb" will be set to "Spot". If this does not happen, the user should click on the "Pull Down" menu under the column "Set Name" and choose "Spot" from the choices provided.

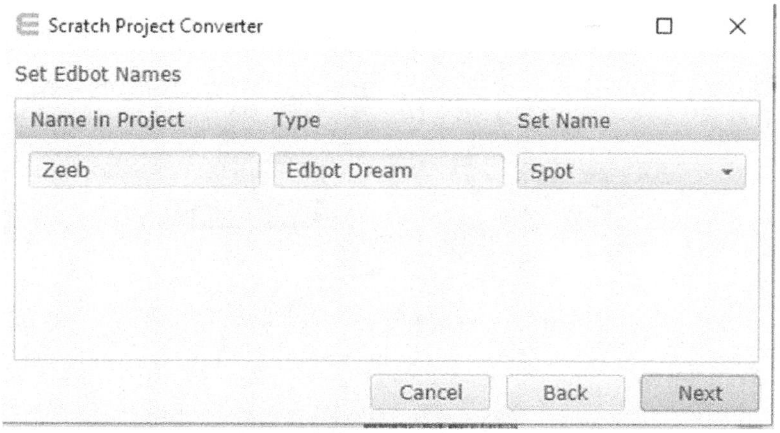

Fig. 2.6 "Set Edbot Names" pop-up window.

Finally, if the user clicks on "Next" as shown in Fig. 2.6, then a final "Summary" window pops up (see Fig. 2.7) where the user can click on "Apply" to affect the actual file conversion.

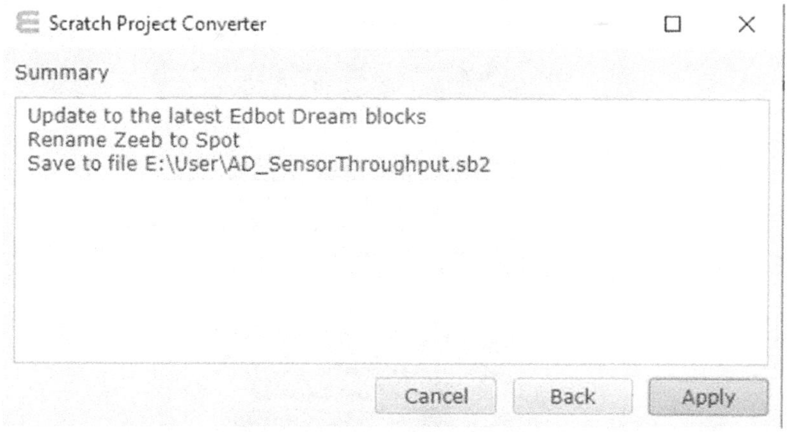

Fig. 2.7 "Summary" window before actual SCRATCH Project Conversion.

As the user can see, only ONE SB2 file can be converted each time that the "Project Converter" is used.

2.3 SCRATCH 2 Blocks Provided with Edbot Software

Edbot provides 13 Stack Blocks and 7 Reporter Blocks which can be accessed under the "More Blocks" item of the SCRIPTS tab (see Fig. 2.8).

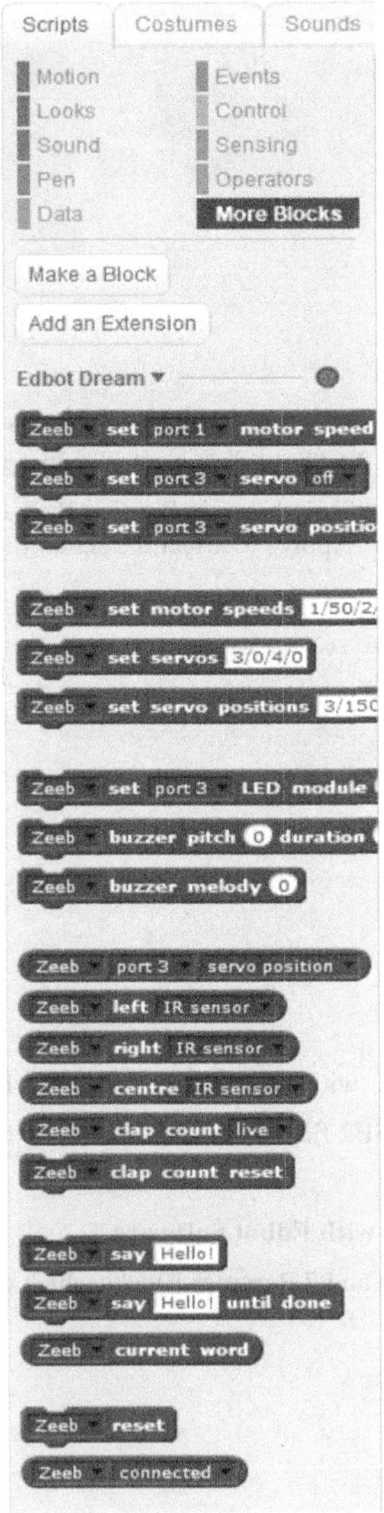

Fig. 2.8 Edbot provides 13 Stack Blocks and 7 Reporter Blocks.

The Edbot website has a support page at http://support.ed.bot/edbot-dream-scratch2.html where the various Blocks are classified as: MOTOR & SERVO, LED & BUZZER, SENSOR, SPEECH and GENERAL.

2.3.1 MOTOR & SERVO Blocks

A quick reminder to the readers regarding the hardware design of the DREAM Controller CM-150 is that Ports 1 and 2 are reserved strictly for Continuous-Turn motors GM-10A (http://www.robotis.us/geared-motor-gm-10a/), while Ports 3 and 4 can be used with Servo motors SM-10 (http://www.robotis.us/servo-motor-sm-10/) which can also function as Continuous-Turn motors. Additionally, Ports 3 and 4 can be used for various Sensors/Actuators such as NIR sensors IRSS-10 or LED modules LM-10. The reader can browse through a list of these sensors at http://www.robotis.us/sensors/.

Edbot offers two types of motor/servo control:

1. Single-Actuator Blocks – i.e. one block is used to control each individual actuator (see Fig. 2.9). There are 3 implementations of these Single-Actuator (SA) blocks: "Set Motor Speed", "Set Servo On/Off" and "Set Servo Position Speed".

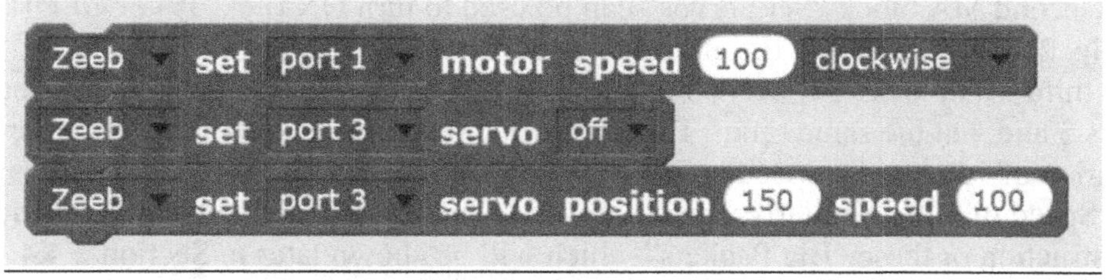

Fig. 2.9 Edbot's SINGLE ACTUATOR (SA) Blocks.

The "Set Motor Speed" block is used to control the speed [0-100] (i.e. STOP-MAXIMUM) and the direction, CLOCKWISE or ANTI-CLOCKWISE (i.e. COUNTER-CLOCKWISE in US English). This block can be used for all four Ports 1 through 4, whereas Ports 1 and 2 are connected to GM-10As and Ports 3 and 4 are connected to SM-10s set to WHEEL mode (i.e. Continuous-Turn mode).

The "Set Servo On/Off" block of course applies to Ports 3 and 4 only, and it is used to activate or deactivate the SM-10 Servos attached to the respective ports. Once a Servo is Off, the user can turn it manually. See Section 2.4.9 for an application of this feature in a video Car Game.

The "Set Servo Position Speed" block is used to set both the Goal Position of the Servo's Horn and the Servo's Speed during the travel to get there. The allowed values for Servo Position are from 0 to 300 (i.e. in degrees – see web page http://support.ed.bot/edbot-dream-scratch2.html for more details). The allowed values for Speed are from 0 to 100, i.e. STOP to MAX as before. The Speed setting for a Servo motor also determines how strong it maintains its Goal Position: for example if Speed is set to 0, the user can rotate the servo's horn manually and thus can use it as a rotational encoder, and its strongest "holding power" is obtained when Speed is set to 100.

2. Multiple-Actuators Blocks – i.e. one block can be used to control multiple actuators simultaneously (i.e. with ONE "big" communication packet via the Edbot software tool) – see Fig. 2.10. **Please note that this feature only works with Edbot V. 4.1.979 or higher, furthermore the CM-150 firmware needs to be V.32 or higher.**

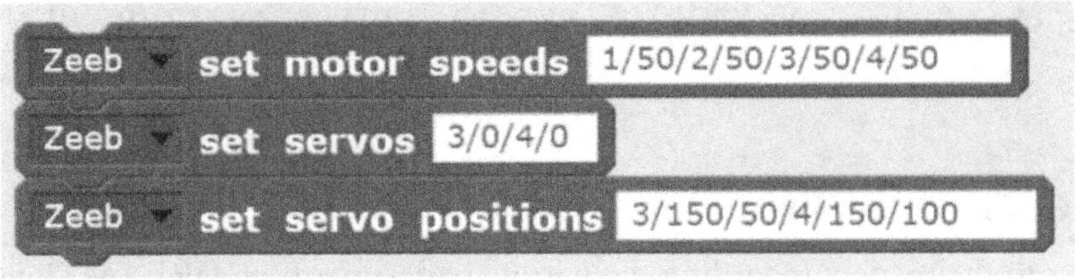

Fig. 2.10 Edbot's MULTIPLE-ACTUATORS Blocks.

In a similar fashion to the Single-Actuator blocks, there are 3 implementations of the Multiple-Actuators (MA) blocks: "Set Motor Speeds", "Set Servos On/Off" and "Set Servo Positions". In the first MA block "Set Motor Speeds", a single message, comprising of 8 **numerical** values separated by the forward-slash symbol "/", is used to send control messages to all 4 ports at the same time. In the example shown in Fig. 2.10, the Speed for Port 1 is set to 50 (message component "1/50"), followed by the message component ("/2/50") which sets the Speed for Port 2 to 50. Next the message component "/3/50" sets Port 3's Speed to 50, and the final component "/4/50" sets Port 4's Speed to 50. There is a way to incorporate Named Variables into the construction of these "Big Packets" which will be shown later in Section 2.4.4.
The second MA block "Set Servos" can be used to turn ON (i.e. "1") or OFF (i.e. "0", as shown in Fig. 2.10) for both Servos on Ports 3 and 4 simultaneously.
The third MA block "Set Servo Positions" sets the Goal Position **and** Speed for Servos on both Ports 3 and 4 at the same time. The example shown in Fig. 2.10 sets Port 3 Servo to the Goal Position of 150 and its Speed at 50, while Port 4 Servo is also set to a Goal Position of 150 but at a Speed of 100. Similarly, there is a way to incorporate Named Variables into the construction of these "Big Packets" which will be shown later in Section 2.4.4.

At this point, the reader probably wonders why these two types of communication/control were implemented and more importantly when/how to use them appropriately. These questions will be answered in Sections 2.4.2 and 2.4.3.

2.3.2 LED & BUZZER Blocks

Fig. 2.11 shows the Edbot blocks used to control the external LED module LM-10 (http://www.robotis.us/led-module-lm-10/) which can be only connected to Ports 3 or 4, and the built-in BUZZER located inside the CM-150 Controller.

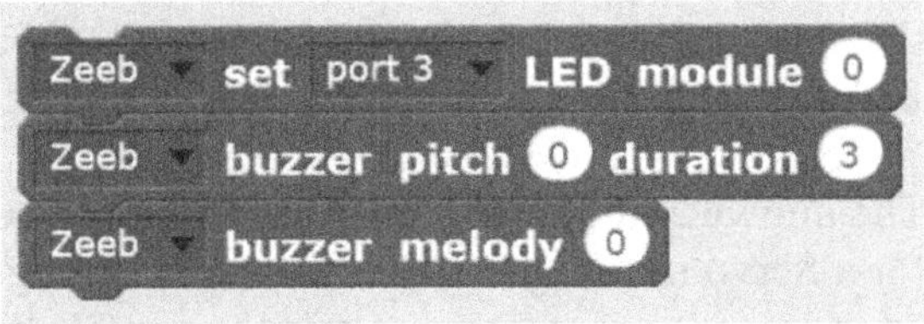

Fig. 2.11 Edbot's LED and BUZZER Blocks.

The "Set LED Module" block requires two inputs: a Port Number (only 3 or 4 are allowed) and a numerical value [0-3] to turn on/off the Orange and Blue LEDs of the LM-10 (details can be found at this link http://support.ed.bot/edbot-dream-scratch2.html).

The "Buzzer Pitch Duration" block also requires two inputs: the Pitch (i.e. Musical Scale) [0-48], and its time duration [0-50], each time unit corresponding to 1/10 of a second (thus a maximum of 5 seconds duration).

The "Buzzer Melody" block can take on a numerical input between 0 and 24, and the chosen Melody will be played for 5 seconds by the CM-150 Controller through its embedded speaker/microphone.

2.3.3 SENSOR Blocks

The first SENSOR block is the "Port Sensor" block which is designed to report on the current Calibrated Values of a variety of Sensors that can be attached to either Ports 3 or 4 (see Fig. 2.12). The definition and range of the numerical values reported by this block vary depending on the sensor type and more details are available at this web page (http://support.ed.bot/edbot-dream-scratch2.html):

- Servo Positions of the SM-10 are reported in degrees [0-300], see diagram on this ROBOTIS link (http://emanual.robotis.com/docs/en/parts/motor/servo_motor/#control-angle).
- For IR Distance Sensors (IRSS-10 and DMS-80), distances between sensor and object are reported in cm, with "100" denoting an "out of range" situation. The working range for the IRSS-10 is from 3 to 30 cm, while the DMS-80 has a longer range from 8 cm to 80 cm.
- The Temperature Sensor (TPS-10) reports in degrees Celsius.
- The Touch Sensor reports "1" if pressed, otherwise "0" is reported.
- The Magnetic Sensor reports "1" if a magnetic object is detected, otherwise it reports "0".
- The user also has the option to just use the "Raw Value" of any sensor [0-1023].

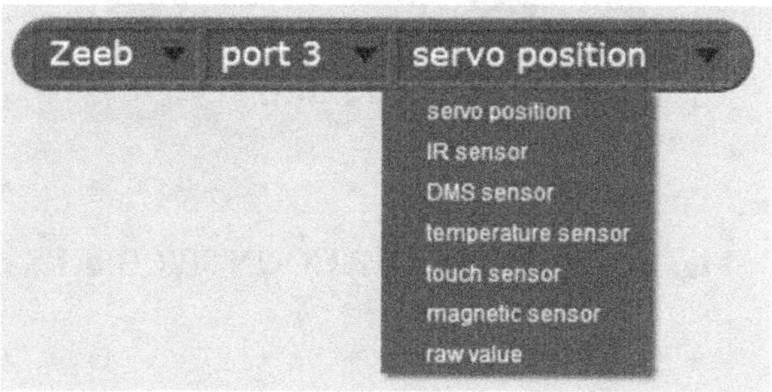

Fig. 2.12 Edbot's PORT SENSOR Blocks.

Next, the built-in NIR sensors, Left-Right-Center, have their own blocks implemented as shown in Fig. 2.13. The user has the choice of using the calibrated values "IR Sensor" (in cm) or the "IR Raw Value" [0~800]. More details are also found at this web link http://support.ed.bot/edbot-dream-scratch2.html.

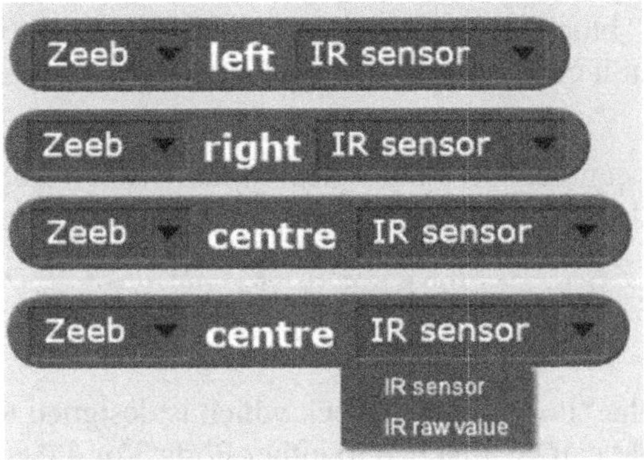

Fig. 2.13 Edbot's Built-In IR SENSOR Blocks.

Lastly, the Clap Counter has a Stack block and a Reporter block implemented as shown in Fig. 2.14. The user should always issue a "Clap Count Reset" command before reading in the "Clap Count", either "Live" or "Last". In practice, the "Last Clap Count" is the most useful data input. For more details, please visit this link http://support.ed.bot/edbot-dream-scratch2.html. Advanced users can also read more on the use of the "Live Clap Count" at this ROBOTIS link (http://emanual.robotis.com/docs/en/software/rplus2/task/#built-in-sensor1-result-of-sound-counter--real-time-sound-counter) – where it is called "Real-Time Sound Counter".

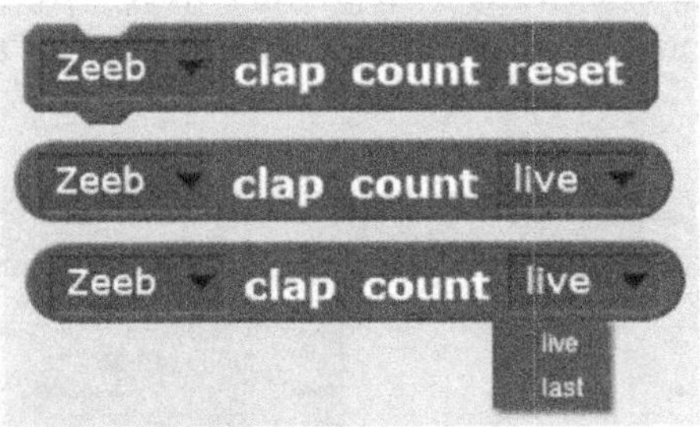

Fig. 2.14 Edbot's CLAP COUNTER Blocks.

2.3.4 SPEECH Blocks

These blocks provide Text-to-Speech capabilities via the Edbot software tool (not via SCRATCH 2). The user can choose the voice of "David" or "Zira" under "Server/Setup" off the main menu bar of the Edbot tool (see Fig. 2.15). The discerning reader may have noticed the check marks for items "Bypass active user" and "Available on network" which will be discussed further in Section 2.6 about Python programming with the Edbot tool.

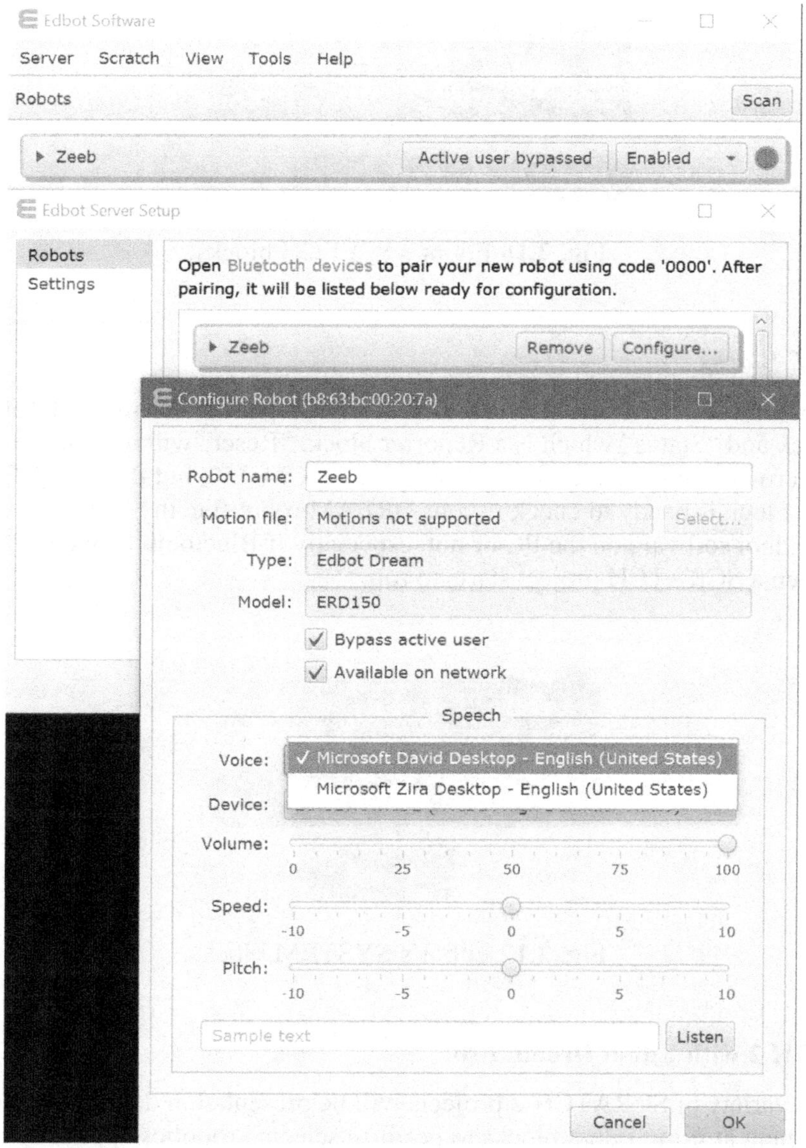

Fig. 2.15 Setting up Edbot (V.5) tool for either "David" or "Zira" voice.

There are 3 speech-related blocks: "Say", "Say Until Done" and "Current Word" (see Fig. 2.16). The difference between "Say" and "Say Until Done" is that "Say" won't wait for the spoken words to be finished before executing the next SCRATCH block, while "Say Until Done" would wait for the speech to finish before continuing to the next SCRATCH block. "Current Word" provides the words being spoken or an "empty text" field if nothing is being spoken, and it can be used to provide visual emphasis during speech operations (see example at this web link http://support.ed.bot/edbot-dream-scratch2.html).

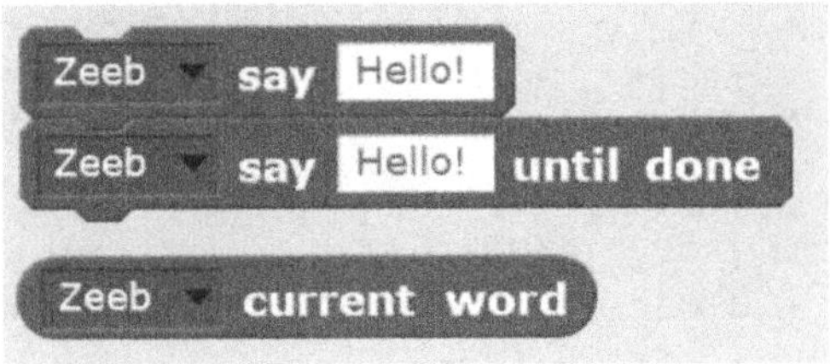

Fig. 2.16 Edbot's SPEECH blocks.

2.3.5 GENERAL Blocks

These GENERAL or SYSTEM blocks relate to the overall operation of the CM-150 Controller: "Reset" which is a Stack block and "Status" which is a Reporter block. "Reset" will turn off all hardware, reset the Last Clap Count to zero and stop all speech operations on the CM-150, but the SCRATCH code will still be active. The "Status" block is handy to check if your DREAM robot (i.e. the CM-150) is "Enabled" or "Connected" to the Edbot software on the PC or not, especially if Bluetooth happens to take a long time to connect properly when a SCRATCH project starts to run.

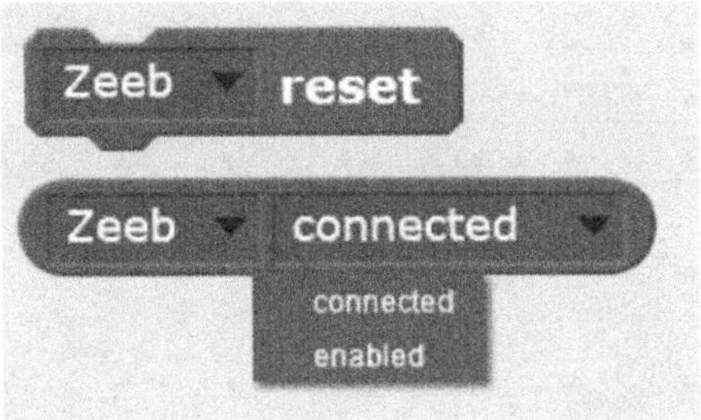

Fig. 2.17 Edbot's SYSTEM blocks.

2.4 Using SCRATCH 2 with Edbot Dream Pro

In this section, a variety of SCRATCH 2 projects will be presented in detail with the goal of helping the reader become proficient at using Edbot blocks to perform selected robotics applications with the ROBOTIS DREAM II system. Where appropriate, the performances between a TASK solution and a SCRATCH-Edbot solution will be compared to enhance the readers appreciation of robotics engineering in general.

Due to a long tool-chain (SCRATCH <-> EDBOT <-> WINDOWS OS <-> BT COMM <-> ROBOT) which can create communication delays, SCRATCH applications usually do not perform very well when sensors data are required at a fast rate for autonomous behaviors such as following a random line track, see an example with a PLAY700 robot (CM-50) at this link (https://www.youtube.com/watch?v=lwpsJu5uRu0). However, the combination of Edbot V.4.1.979 (or newer) and Firmware V.32 (or newer) for the CM-150 provides much performance improvement as demonstrated in the following sub-sections.

2.4.1 Sensor Throughput Performance

The project "AD_SensorThroughput.sb2" is shown in Fig. 2.18 whereas:

1. Line 1 waits for Key "1" to be pushed on the PC's keyboard to start the data acquisition and display procedure.
2. Line 2 provides an important step which is to wait for Bluetooth connections to be established between the SCRATCH IDE (via Edbot Tool) and the robot Zeeb (i.e. the CM-150 controller).
3. Lines 3 and 4 work together to let the CM-150 play out Buzzer Melody 3, before going on to the Endless Loop which performs the actual data acquisition and display functions.
4. Line 5 starts the Forever Loop which includes 5 Action blocks (Lines 6 through 10).
5. Line 6 uses a Reporter block to obtain Zeeb's "Left IR Raw Value" and saves it in Parameter "IR_Left" which is then displayed in the Stage area (see Fig. 2.19).
6. A similar process to Line 6 is then applied respectively to 4 more Sensors/Parameters: "IR_Center", "IR_Right", "IR_Port3" and "IR_Port4" (Lines 7 to 10).

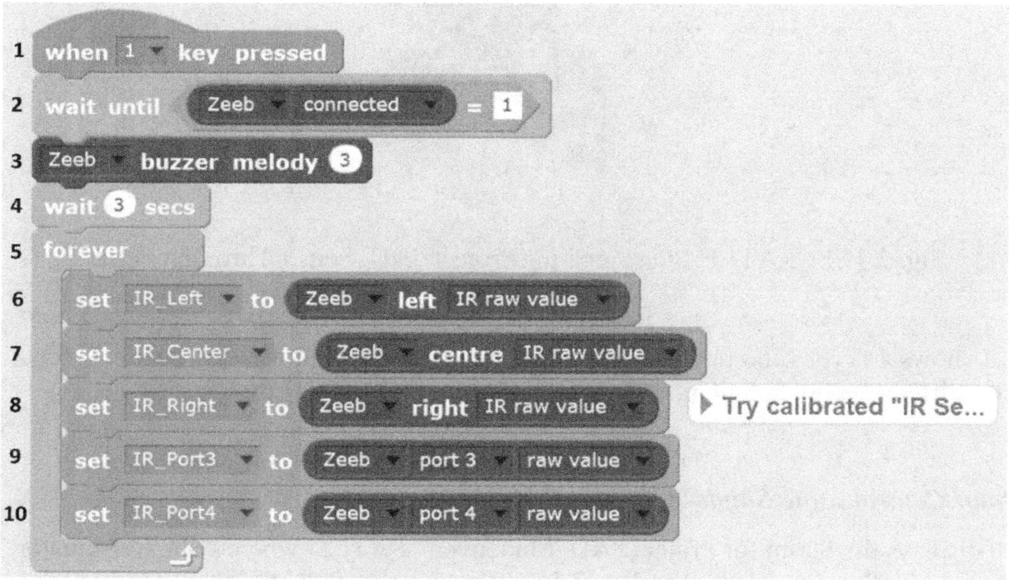

Fig. 2.18 SCRATCH Script for Project "AD_SensorThroughput.SB2".

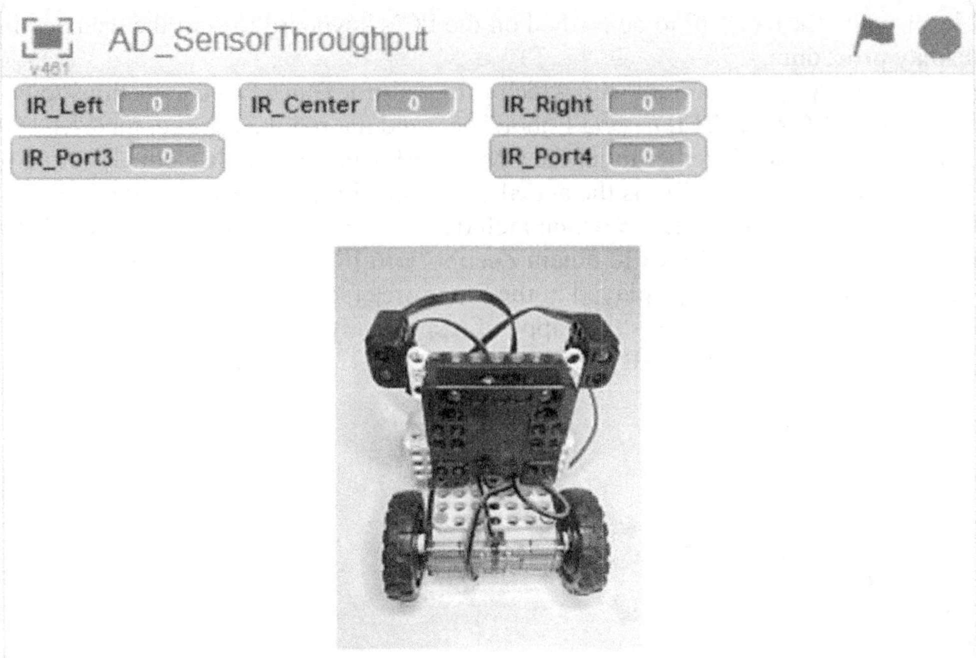

Fig. 2.19 SCRATCH Stage area for Project "AD_SensorThroughput.SB2".

Video 2.1 shows a fairly good rate of overall data transmission, and we shall see if this data throughput is adequate for different robot autonomous behaviors in later SCRATCH projects.

2.4.2 Motor Control using Single-Actuator (SA) Blocks

Fig. 2.20 displays the Script for Project "AD_Maneuvers_SA.sb2" where "Single-Actuator" blocks are used to send sequential commands to Port 1 then Port 2 in order to make the robot first go forward and straight at 50% power for 1 second (Lines 5-7), next stop both motors (i.e. the robot) for 2 seconds (Lines 8-10), then make the robot go backward for 1 second at 50% power (Lines 11-13), and finally stop the robot for good (Lines 14-15). Just from common sense, the user would expect the robot to go forward and then backward in a "straight" line, but Part 1 of Video 2.2 shows that the robot did not do so at run time: there was a jerk on the robot's left at the beginning of the forward motion, a jerk on the robot's right at the end of the forward motion, and then once again, a jerk on the robot's left at the start of the backward motion, then a final jerk on the robot's right at its final stop. Some observant readers may already notice and conclude that Port 1 Motor (on robot's left) always started (or ended) its motion before the Port 2 Motor (on robot's right). The question is then why?

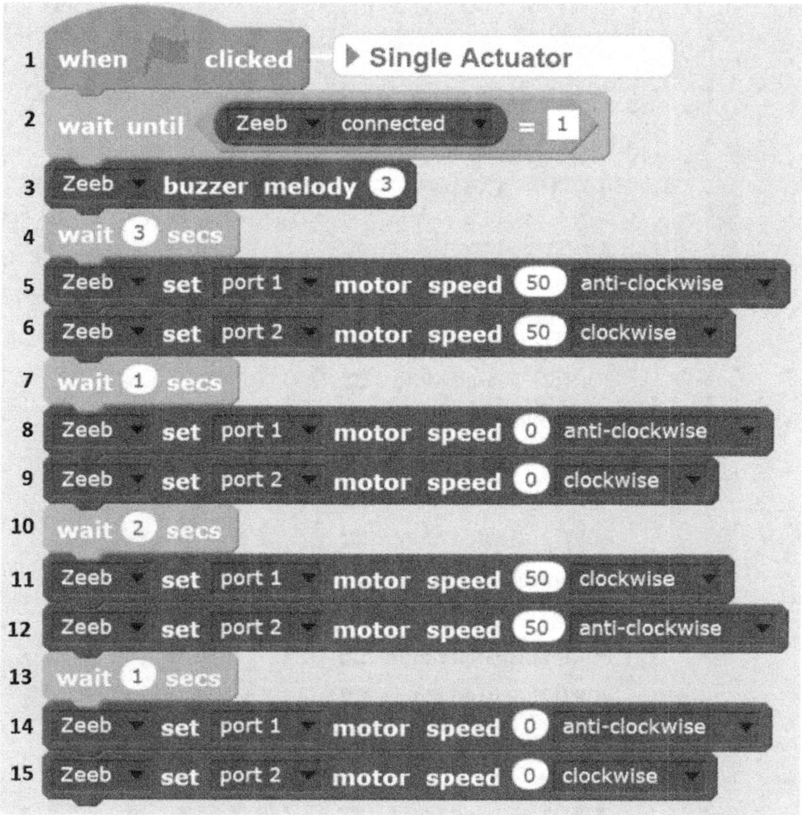

Fig. 2.20 SCRATCH Script for Project "AD_Maneuvers_SA.SB2".

But before answering that question, the reader needs to view Part 2 of Video 2.2 next. Part 2 shows the run-time performance of the equivalent TASK code called "AD_Maneuvers_SA.tskx" as listed in Fig. 2.21.

```
3  START PROGRAM
4  {
5      Buzzer Timer = Melody Time
6      Buzzer Index = Melody No.3 (3)
7      WAIT WHILE ( Buzzer Timer > 0.0sec )
8  
9      Result of Sound Counter = 0
10     WAIT WHILE ( Result of Sound Counter == 0 )
11     PORT[1]:Geared Motor = CCW:511 (49.95%)
12     PORT[2]:Geared Motor = CW:511 (49.95%)
13     High-resolution Timer = 1.000sec
14     WAIT WHILE ( High-resolution Timer > 0.0sec )
15     PORT[1]:Geared Motor = CW:0 (0.00%)
16     PORT[2]:Geared Motor = CCW:0 (0.00%)
17     High-resolution Timer = 2.000sec
18     WAIT WHILE ( High-resolution Timer > 0.0sec )
19     PORT[1]:Geared Motor = CW:511 (49.95%)
20     PORT[2]:Geared Motor = CCW:511 (49.95%)
21     High-resolution Timer = 1.000sec
22     WAIT WHILE ( High-resolution Timer > 0.0sec )
23     PORT[1]:Geared Motor = CCW:0 (0.00%)
24     PORT[2]:Geared Motor = CW:0 (0.00%)
25 }
```

Fig. 2.21 TASK Code "AD_Maneuvers_SA.tskx".

Comparing Fig. 2.21 to Fig. 2.20, the reader should note that the logical sequence of the main action commands to the robot are the same:

- Lines 3-4 of Fig. 2.20 corresponds to Lines 5-7 of Fig. 2.21 (i.e. play Melody 3).
- Lines 5-7 of Fig. 2.20 corresponds to Lines 11-14 of Fig. 2.21 (forward at 50% power for 1s).
- Lines 8-10 of Fig. 2.20 corresponds to Lines 15-18 of Fig. 2.21 (stop for 2s).
- Lines 11-13 of Fig. 2.20 corresponds to Lines 19-22 of Fig. 2.21 (backward at 50% for 1 s).
- Lines 14-15 of Fig. 2.20 corresponds to Lines 23-24 of Fig. 2.21 (final stop).

However, the TASK code did not result in the jerking motions exhibited by the SCRATCH code, although both codes used "Single-Actuator" types of command to activate Ports 1 and 2. The difference lies in the location where the "EXECUTING" code resides with respect to the CM-150:

- For the combo SCRATCH/Edbot, the executing code resides inside the SCRATCH IDE (i.e. on the Windows PC) where it gets interpreted block by block and sent to the Edbot tool which then translates them into appropriate commands that the CM-150 would understand and execute, and all that communication traffic is done via serial Bluetooth at only 57.6 kbps. Obviously, this is a time-consuming approach as compared to the TASK approach.
- When using the TASK tool, the binary machine code equivalent to the commands listed in the TASK code is already downloaded to the local memory of the CM-150 controller. Thus, at

execution time, even though there may be a theoretical time delay between Single-Actuator commands (such as Lines 11-12 in Fig. 2.21), that time delay is so small that it would not result in a "visible" physical delay as exhibited by the SCRATCH/Edbot approach.

This test shows that we can expect some systemic delays in the autonomous SCRATCH projects that would require fast sensor/actuator synchronization as demonstrated in later sections.

2.4.3 Motor Control using Multiple-Actuators (MA) Blocks

To mitigate the timing delays for activating and deactivating motors attached to Ports 1 and 2, one approach is to send a "big data packet" containing commands to both Ports using a single SCRATCH block and once that big packet is received in the local memory of the CM-150, we'll let the CM-150 deliver the actual commands to Ports 1 and 2 as normal. But now, as the command sending process is from the CM-150's local memory, it will have minimal time delays as with the binary TASK code - as described in Section 2.4.2.

Fig. 2.22 displays the Script for Project "AD_Maneuvers_MA.sb2" where "Multiple-Actuators" (MA) blocks are used to send activation and deactivation commands to Ports 1 and 2 "simultaneously". The same sequence of commands are issued to make the robot first go forward and straight at 50% power for 1 second (Lines 5-6), next stop both motors (i.e. the robot) for 2 seconds (Lines 7-8), then make the robot go backward for 1 second at 50% power (Lines 9-10), and finally stop the robot for good (Line 11). Please note that a positive value for the motor speed produces a counter-clockwise rotation of the motor, and vice-versa for a negative value of the motor speed (see Lines 5 and 9).

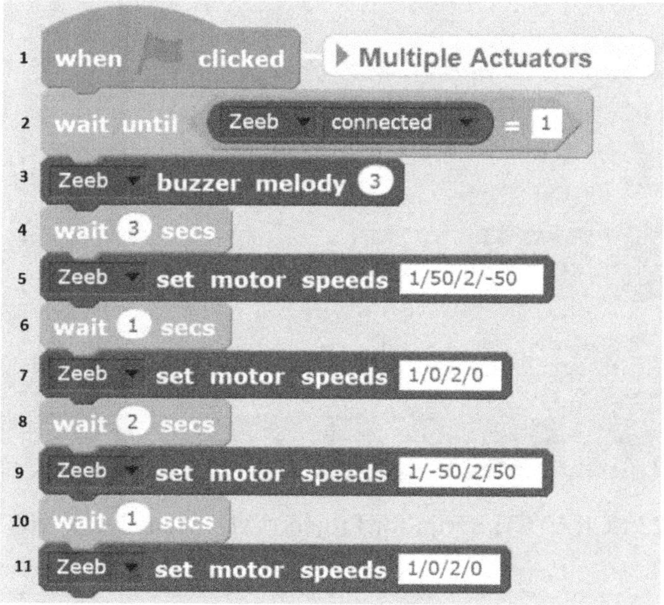

Fig. 2.22 SCRATCH Script for Project "AD_Maneuvers_MA.SB2".

Video 2.3 shows that the activation/deactivation timing delays of the motors on Ports 1 and 2 are no longer perceptible. **From now on, we will be using MA blocks in our SCRATCH projects whenever it is appropriate and efficient to do so**. Please note that the MA feature is not available if the user is using the ROBOTIS R+SCRATCH/SCRATCH 2 tool chain as demonstrated in the author's first DREAM book (Thai 2018).

2.4.4 Variable Speed Motor Control with MA Blocks

Reviewing Fig. 2.22, the reader may have already noticed that the motor speed values were specified as numerical constants (50 or -50) in Lines 5 and 9, but sometimes a Variable Parameter (with such name as "Speed") may have to be used, thus somehow we'll have to incorporate the Numerical Value represented by the String "Speed" into the creation of those "Big Data Packets" used in Lines 5 and 9. The SCRATCH's JOINT Operator can be used for this purpose as shown in Fig. 2.23 for the next SCRATCH project named "AD_Maneuvers_MAV.SB2". In this project, the reader needs to create a User Block (i.e. User Function) named "Move", using the "Make a Block" tool, available under the option "More Blocks" of the SCRIPTS Tab (see Fig. 2.24). If the reader is not familiar with this procedure, please watch the first part of Video 2.4 for a step-by-step tutorial on how to create and define the blocks used in this User Block named "Move".

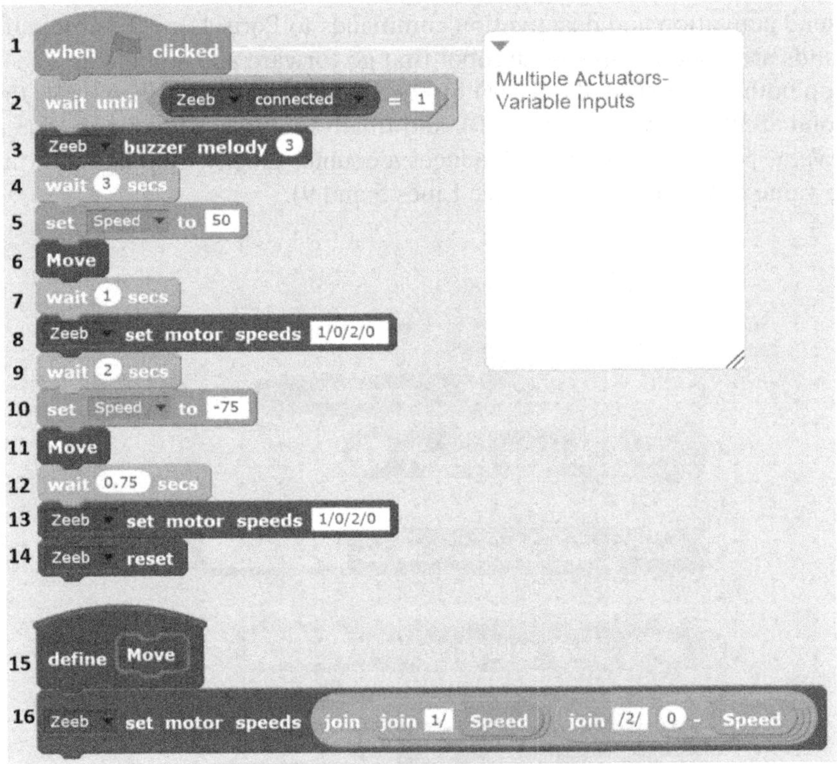

Fig. 2.23 SCRATCH Script for Project "AD_Maneuvers_MAV.SB2".

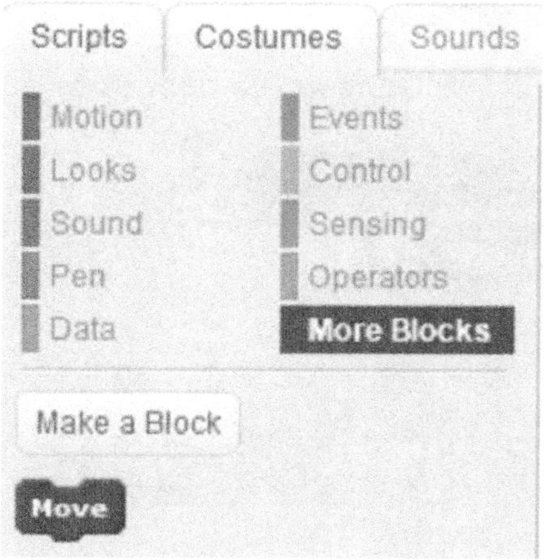

Fig. 2.24 Accessing Tool named "Make a Block" to create a User Block (i.e. User Function).

In Project "AD_Maneuvers_MAV.SB2" (see Fig. 2.23), the author's goal was first to send the robot forward at 50% power to the motors at Ports 1 and 2 (Lines 5-7), and then to send the robot backward at 75% motor power (Lines 10-12). A Variable Parameter named "Speed" was used to contain at first a numerical value of "50" (Line 5) which was changed to "-75" after (Line 10). A User Block named "Move" (Lines 6 and 11) was used to make the Script more efficient as the complex block using the JOINT operator had to be defined only once (Lines 15-16). Fig. 2.25 illustrates the actual String "1/Speed/2/-Speed" to be created and its implementation using several JOIN operators and one SUBSTRACTION operator.

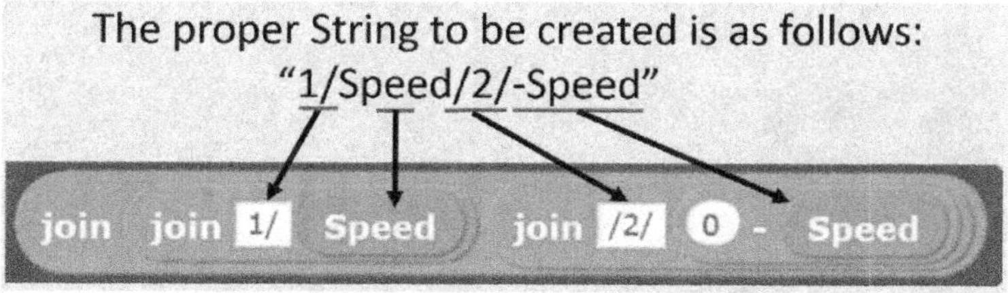

Fig. 2.25 How to create the string "1/Speed/2/-Speed" using JOIN & SUBSTRACTION operators.

Part 2 of Video 2.4 shows that the robot performs the maneuvers as expected.

2.4.5 Avoider (R+SCRATCH vs. Edbot)

The materials in this Section 2.4.5 can be compared to Section 4.5 of the author's first DREAM book (Thai 2018) where the tool chain SCRATCH 2/R+SCRATCH was used, and it could only support Single-Actuator blocks.

Fig 2.26 shows the main script for "AD_Avoider-IF_ELSE_IF_1.SB2" where the main algorithm is controlled by a Forever Loop (Lines 7-25):

- For each loop, Lines 8-9 collect and save current values for Left and Right IR Raw Values into Variables IR_Left and IR_Right respectively.

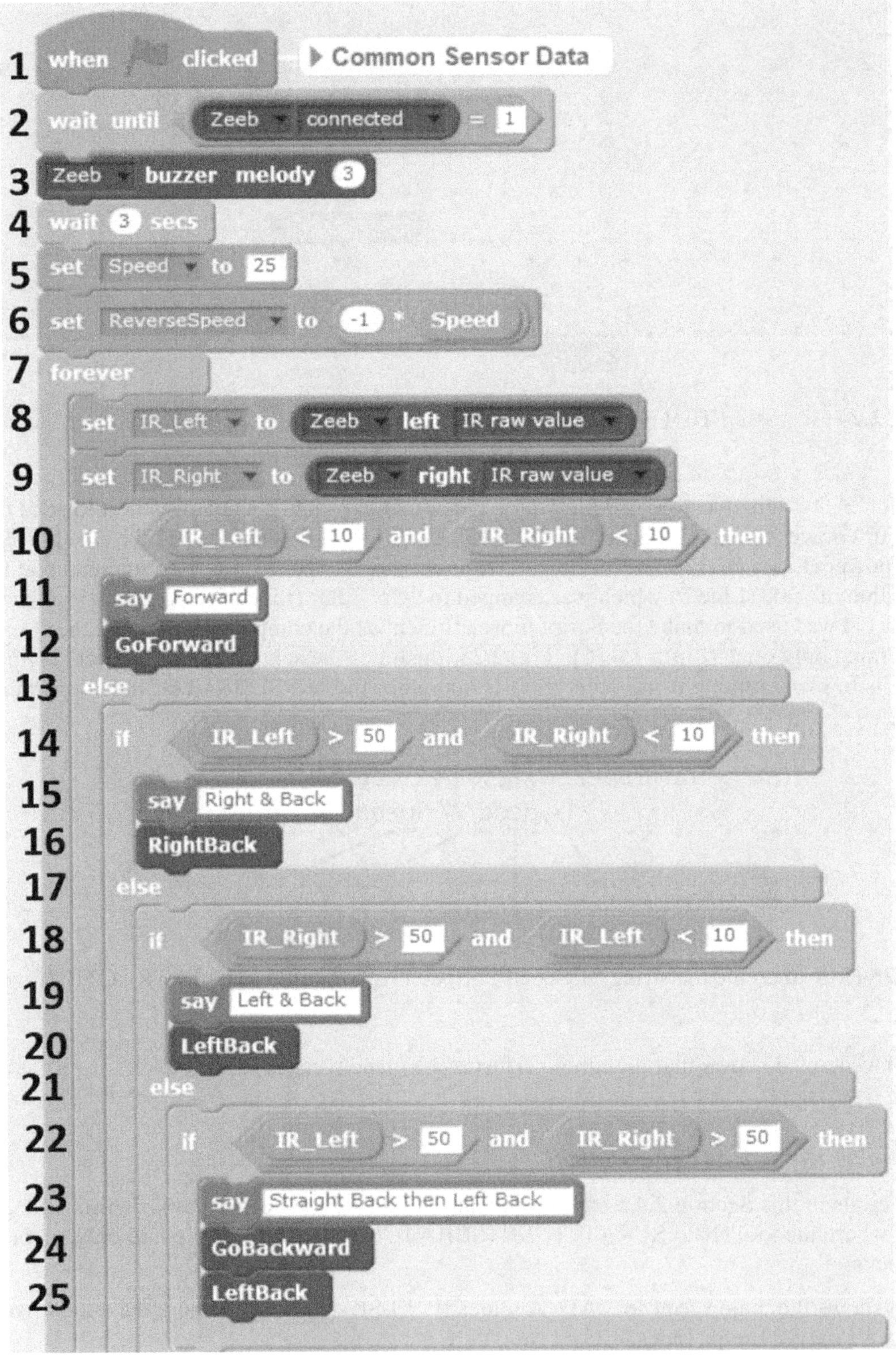

Fig. 2.26 Main Script for Project "AD_Avoider-IF_ELSE_IF_1.SB2".

- Next, a big IF-ELSE-IF structure (Lines 10-25) is used to decide on 4 possible Actions (GoForward, RightBack, LeftBack, or GoBackward+Leftback) based on the appropriate conditions of the parameters "IR_Left" and "IR_Right" (Lines 10, 14, 18 or 22). The overall goal for this IF-ELSE-IF structure is to help the robot avoid obstacles that are getting too close to the built-in Left and Right IR Sensors, i.e. when IR_Left or IR_Right values are greater than 50 (Lines 14, 18 and 22). If IR_Left and IR_Right are less than 10 (Line 10), it is OK for the robot to go forward.

Fig. 2.27 lists the definitions of the User Blocks "GoBackward", "RightBack", "LeftBack" and "GoForward" using Edbot's Multiple-Actuators blocks.

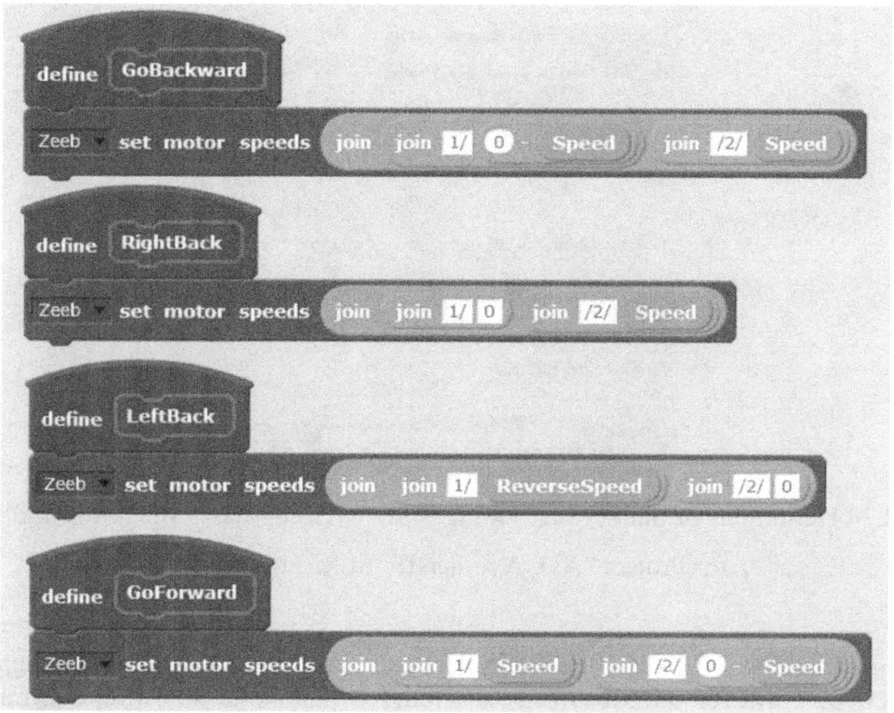

Fig. 2.27 Definitions of User Blocks "GoBackward", "RightBack", "LeftBack" and "GoForward" for Project "AD_Avoider-IF_ELSE_IF_1.SB2".

Fig. 2.28 shows 3 independent SubScripts used to make the robot "Going Slow" (via the "1" Key) or "Going Fast" (via the "2" Key), or to "Stop and Reset" the robot temporarily (via the "Space" Bar).

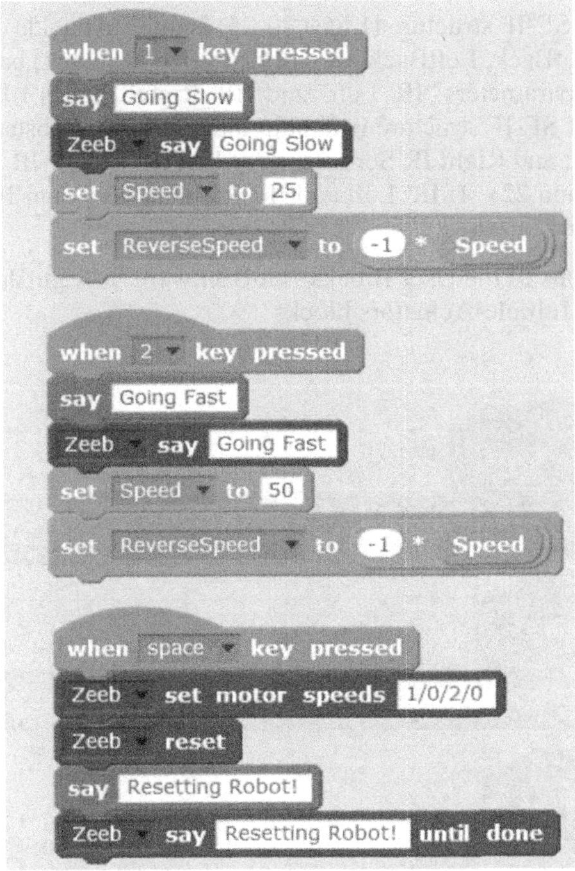

Fig. 2.28 Definitions of SubScripts "Going Slow", "Going Fast" and "Stop and Reset" for Project "AD_Avoider-IF_ELSE_IF_1.SB2".

Video 2.5 shows that this SCRATCH/Edbot Project makes the robot responsive enough to avoid obstacles in front of it. However, once the obstacle is removed, the robot was still performing avoiding maneuvers for a few more cycles before going forward again as it should. Reviewing Video 2.5 more carefully, the reader would notice that the message "Forward" was displayed quite promptly in the SCRATCH "Stage" area upon the removal of the frontal obstacle. This means that the Edbot tool can handle the "Sensor Data" flow from the CM-150 to the SCRATCH 2 IDE fast enough for autonomous sensor-intensive behaviors. However, Edbot somehow handles the "Motor Commands" flow from the SCRATCH 2 IDE to the CM-150 more slowly, perhaps a larger buffer was used for the SCRATCH to CM-150 data flow? As the author does not have knowledge of the inner workings of the Edbot tool, he cannot say anything for sure.

Next, Project "AD_Avoider-IF_ELSE_IF_2.SB2" was created with the goal of using the "Current NIR Sensor Data" directly as arguments inside the various conditional statements IF and ELSE-IF (see Lines 8, 12, 16 and 20 of Fig. 2.29) with the hope that it would mitigate the previous data flow problems.

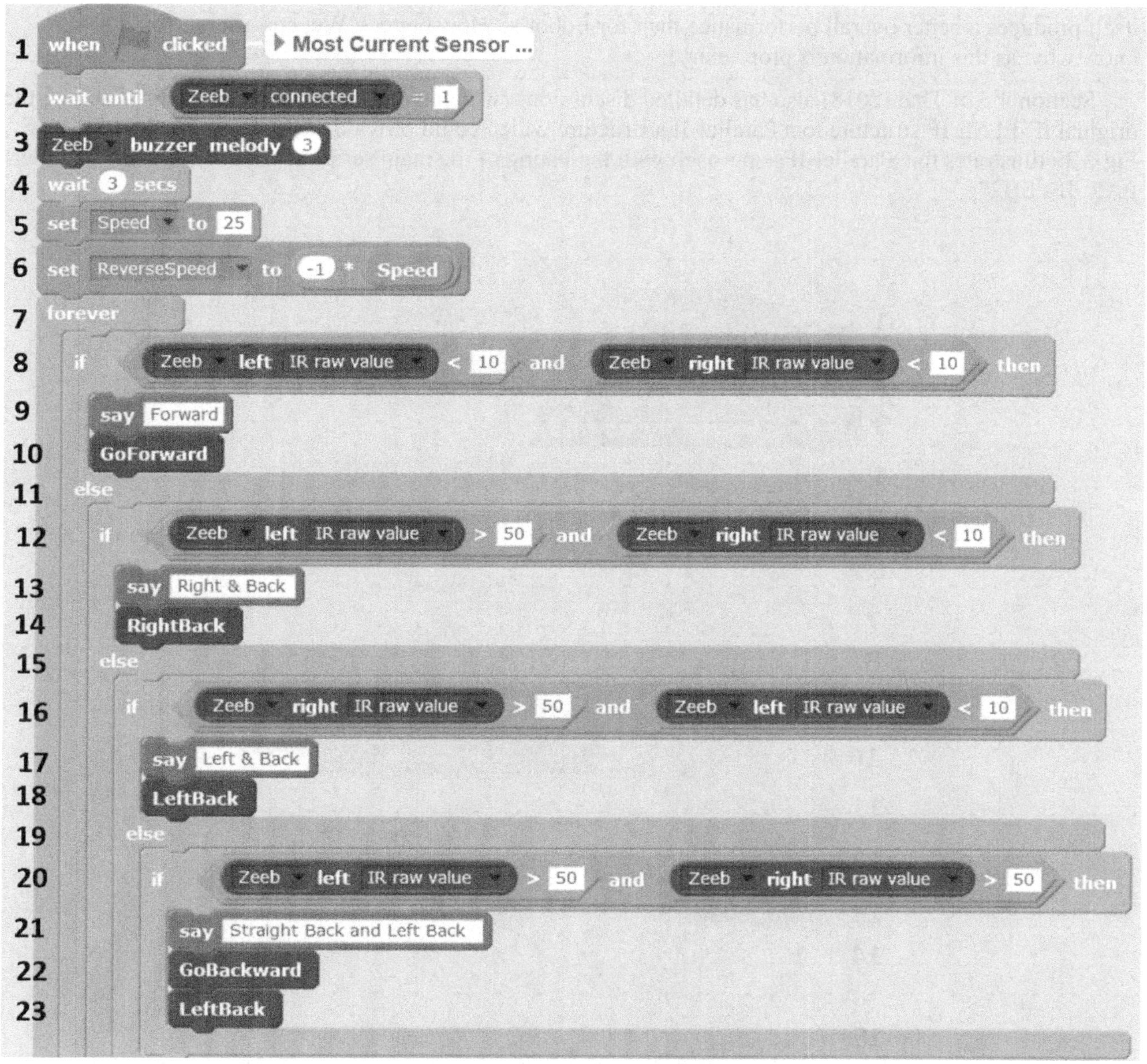

Fig. 2.29 Main Script for Project "AD_Avoider-IF_ELSE_IF_2.SB2".

Video 2.6 shows that the same problems remain.

In Section 4.5 of the author's first DREAM book (Thai 2018), a R+SCRATCH/SCRATCH 2 project was created to handle the same obstacle-avoiding behavior using an IF-ELSE-IF structure and using Single-Actuator blocks (the file "AD-Avoider-2.SB2" included with this book) and the reader can watch this YouTube video at https://www.youtube.com/watch?v=KImcweE8KhQ, which shows that the R+SCRATCH/SCRATCH 2 solution was quite adequate for autonomous behaviors with the CM-150. This means that R+SCRATCH's approach/code for handling sensor data and actuator commands in and out of itself produces a better overall performance than for Edbot's. How better? We, end-users, would never know why, as this information is proprietary!

Section 4.5 of Thai (2018) also has detailed discussions (which won't be repeated here) for changing the original IF-ELSE-IF structure to a Parallel-IFs structure which could provide a better runtime performance. Fig 2.30 illustrates this Parallel-IFs approach with the listing of the main script of the Project "AD_Avoider-PAR_IFs.SB2".

Fig. 2.30 Main Script for Project "AD_Avoider-PAR_IFs.SB2".

Lines 11-14 and Lines 15-18 denote the two Parallel IFs used. The reader should also note that the Conditional Statements used are simpler (i.e. quicker to execute at runtime) than the ones used by the IF-ELSE-IF structures as shown in Figs. 2.26 and 2.29.

The Parallel IFs algorithm also requires a Time Delay (Line 7) to let the designated actions play out (Lines 14 and 18). The user can tune the parameter "DelayTime" to obtain the optimal overall response for his or her system. For the author's robot, 0.5 second was about the best value to use, values less or more than 0.5 second made this robot less responsive when the frontal obstacle was abruptly removed.

Finally, the Parallel-IFs approach always has a default action which is "GoForward" in this case and it is always laid out as the last action after the Parallel IFs (Lines 19-20).

Video 2.7 shows that there is much improvement in responsiveness to random obstacle movements over the IF-ELSE-IF structure.

Lastly, Section 4.5 of Thai (2018) also discussed the use of a Parallel-LOOPs structure which also could provide a better runtime performance. Fig 2.31 illustrates this Parallel-LOOPs approach with the listing of the main script of the Project "AD_Avoider-PAR_Loops.SB2".

Fig. 2.31 Main Script for Project "AD_Avoider-PAR_Loops.SB2".

Lines 9-12 and Lines 13-16 denote the two Parallel REPEAT_UNTIL loops used. The reader should also note that the Conditional Statements used are also simpler (i.e. quicker to execute at runtime) than the ones used by the IF-ELSE-IF structures as shown in Figs. 2.26 and 2.29. Furthermore, "Current" Sensor Data (i.e. Reporter blocks) will need to be used for Loops to work properly (Lines 9 and 13).

Similarly, the Parallel Loops algorithm requires a Time Delay (Line 7) to let the designated actions play out (Lines 12 and 16). Again, the user can tune Parameter "DelayTime" to obtain the optimal overall response for his or her system. For the author's robot, 0.5 second was also about the best value to use, values less or more than 0.5 second made this robot less responsive when the frontal obstacle was abruptly removed.

The Parallel-LOOPs approach also needs the default action ("GoForward") to be laid out as the last action after the Parallel Loops (Lines 17-18).

Finally, the Parallel-LOOPs approach has the distinction of using the least number of blocks than the other two approaches.

Video 2.8 also shows that there is much improvement in responsiveness to random obstacle movements over the IF-ELSE-IF structure.

2.4.6 Yin-Yang Track Follower

In a way, this Section 2.4.6 reverses the logical perspective of the Section 2.4.5 as the DREAM robot will now use its NIR sensors to try to follow/track a frontal object.

But before getting on the Yin-Yang Track Project, let's first start simple with Project "AD_Follower-PAR_IF1.sb2" which Main Script is described in Fig. 2.32. The main loop starts by collecting the latest NIR Sensor readings and save them in Parameters IR_Left and IR_Right (Lines 9-10).

The two Parallel IFs are denoted by Lines 11-14 and 15-18 and they need a DelayTime of 0.1 sec to perform their moves per loop cycle. The default move is to stop the robot (Lines 19-20).

Video 2.9 shows that there is a delay in starting the tracking process but that there is usually no overshoot. The high-speed setting seems to work better that the slow-speed setting (probably not enough torque into the motors).

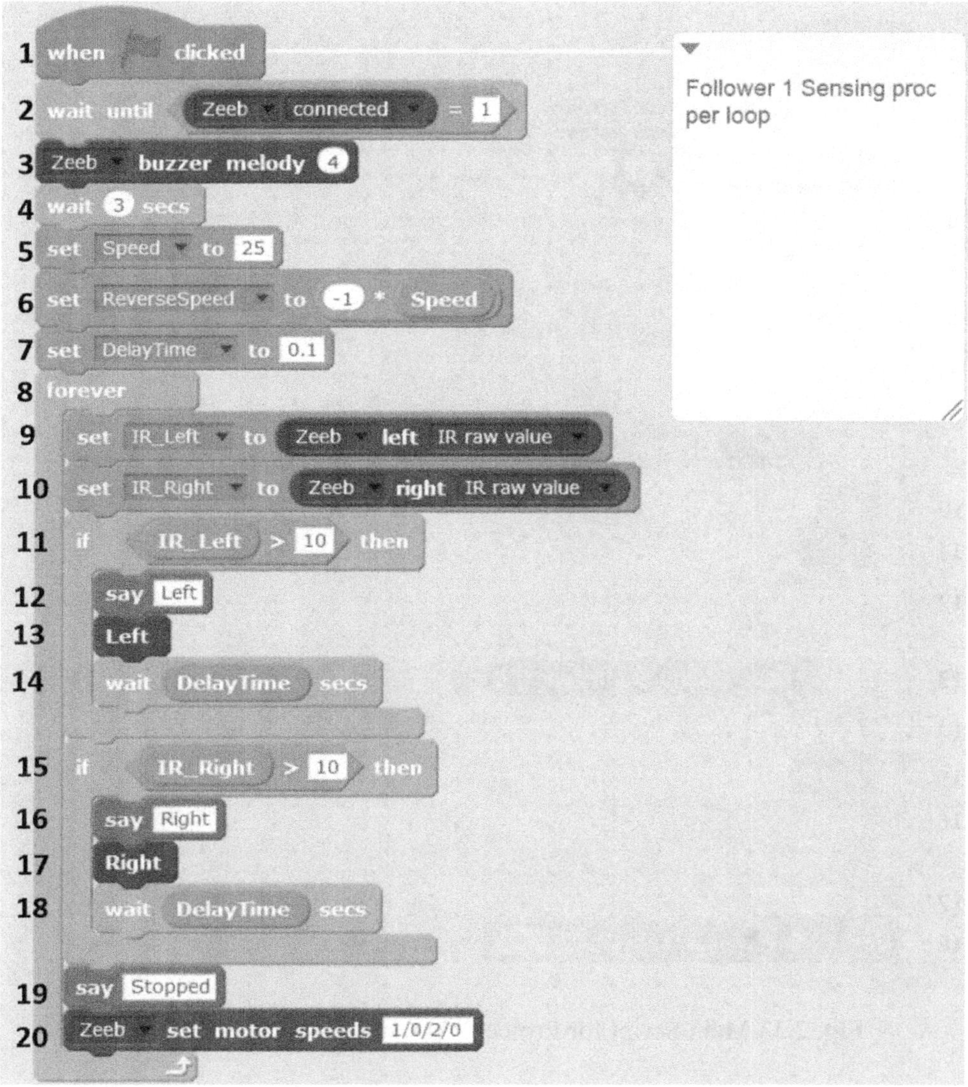

Fig. 2.32 Main Script for Project "AD_Follower-PAR_IF1.SB2".

The next Project "AD_Follower-PAR_IF2.sb2" which Main Script is described in Fig. 2.33. The main loop starts with the first IF structure (Lines 9-12) using a Reporter block directly in its Conditional Expression (Line 9) and using a DelayTime of 0.1 sec (Lines 7 and 12). The second IF is similarly denoted by Lines 13-16. The default move is to stop the robot (Lines 17-18). This Project uses two blocks less than the previous one, however Video 2.10 shows that this algorithm tends to overshoot the object, so it tends to swing back and forth before settling down.

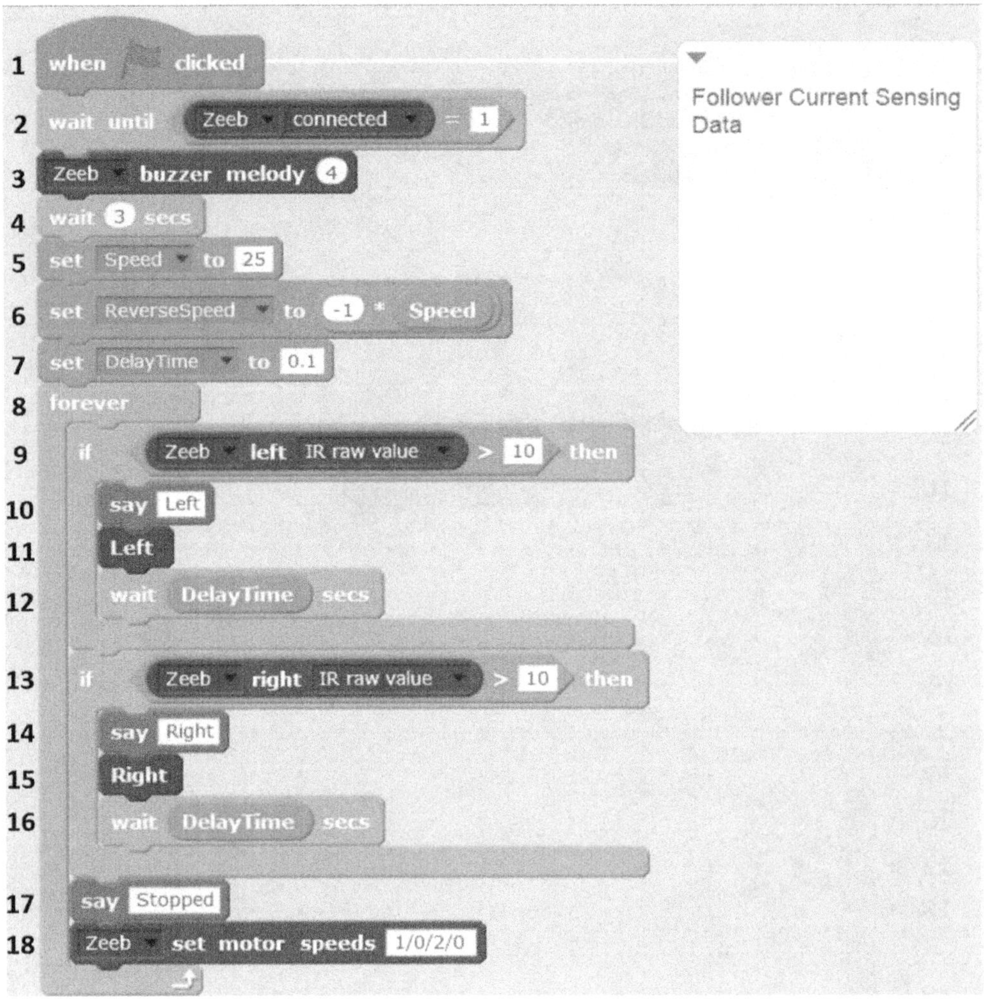

Fig. 2.33 Main Script for Project "AD_Follower-PAR_IF2.SB2".

The third Follower Project uses Parallel Loops instead and its Main Script is illustrated in Fig. 2.34. The main loop contains 2 REPEAT UNTIL Loops (Lines 9-11 and Lines 12-14) using a Reporter block directly in its Conditional Expression (Lines 9 and 12) and do not need a DelayTime. The default move is still to stop the robot (Lines 15-16). This Project uses the least number of blocks among the 3 approaches discussed, however Video 2.11 shows that this algorithm tends to overshoot even worse that the PAR_IFs_2 version (Fig. 2.32), so this is not a recommended approach to object following or target tracking.

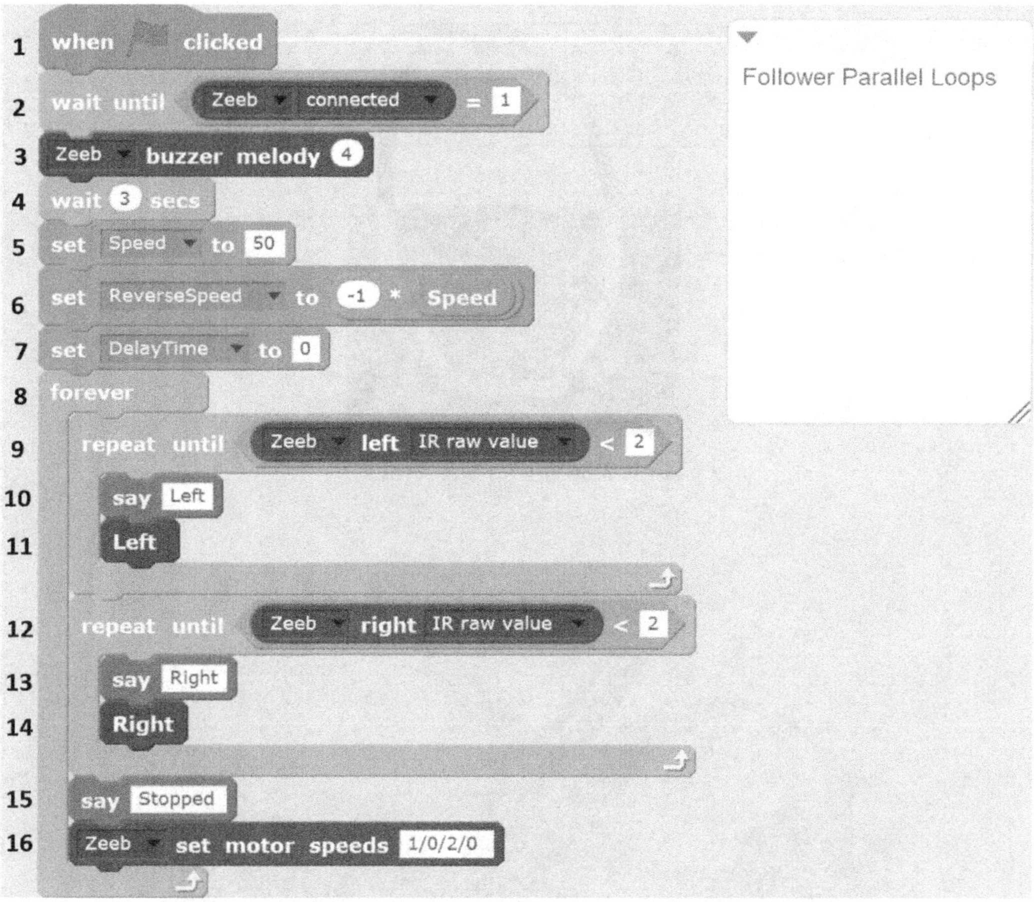

Fig. 2.34 Main Script for Project "AD_Follower-PAR_Loops.SB2".

We are now ready to tackle the Yin-Yang Track Following Project. The Yin-Yang Track gets its name from the feature that the track switches from Black to White and vice-versa for half of its total layout area (see Fig. 2.35). Analyzing this Yin-Yang track further, the author realized that more NIR sensors are needed as the background characteristics had to be known in addition to the usual track characteristics to help steer the robot onto the "right track". Thus, two NIR Sensors IRSS-10 were connected to Ports 3 and 4 and mounted to the front of the robot as shown in Fig 2.36.

Section 4.7 of Thai (2018) discussed the details of how to track an arbitrary black track on a white background and the essential algorithm reduced to two mutually exclusive condition/action pairs and one default action:

1. If the Left IR Sensor sees Black and Right IR Sensor sees White, then the robot needs to turn Left to get back to the black line.
2. Else If the Right IR Sensor sees Black and the Left IR Sensor sees White, then the robot needs to turn Right to get back to the black line.
3. Else the robot needs to go Forward (and the robot needs to start its trek straddling on the black line also).

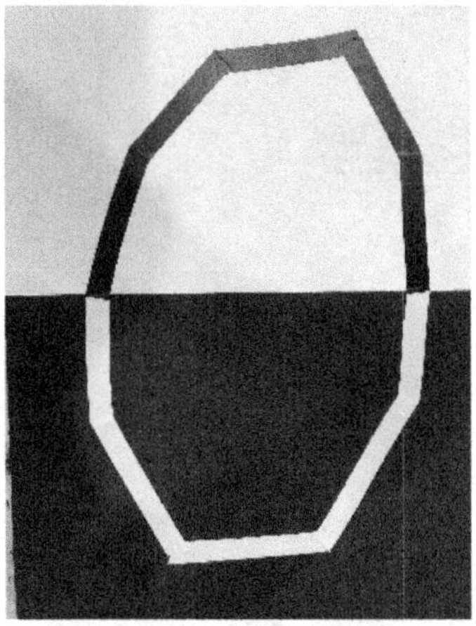

Fig. 2.35 Yin-Yang Track.

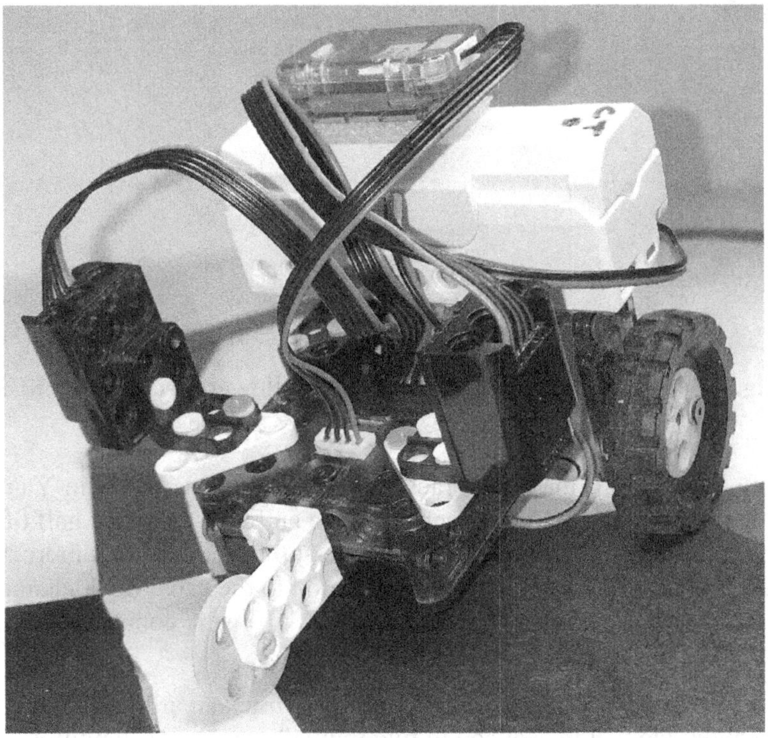

Fig. 2.36 Yin-Yang Track Following Robot.

Adapting this algorithm to the Yin-Yang track results in 4 mutually exclusive condition/action pairs and 1 default action (because additionally the robot needs to know which background it is on – black or white):

1. If the Left IR Sensor sees Black and Ports 3 & 4 IR Sensors see White, then the robot needs to turn Left to get back to the black line.
2. Else If the Right IR Sensor sees Black and Ports 3 & 4 IR Sensors see White, then the robot needs to turn Right to get back to the black line.

3. Else If the Left IR Sensor sees White and Ports 3 & 4 IR Sensors see Black, then the robot needs to turn Left to get back to the white line.
4. Else If the Right IR Sensor sees White and Ports 3 & 4 IR Sensors see Black, then the robot needs to turn Right to get back to the white line.
5. Else the robot needs to go Forward (and similarly the robot needs to start its trek straddling on the black/white line).

Fig. 2.37 illustrates the implementation of the above algorithm into a SCRATCH 2 Forever Loop where:

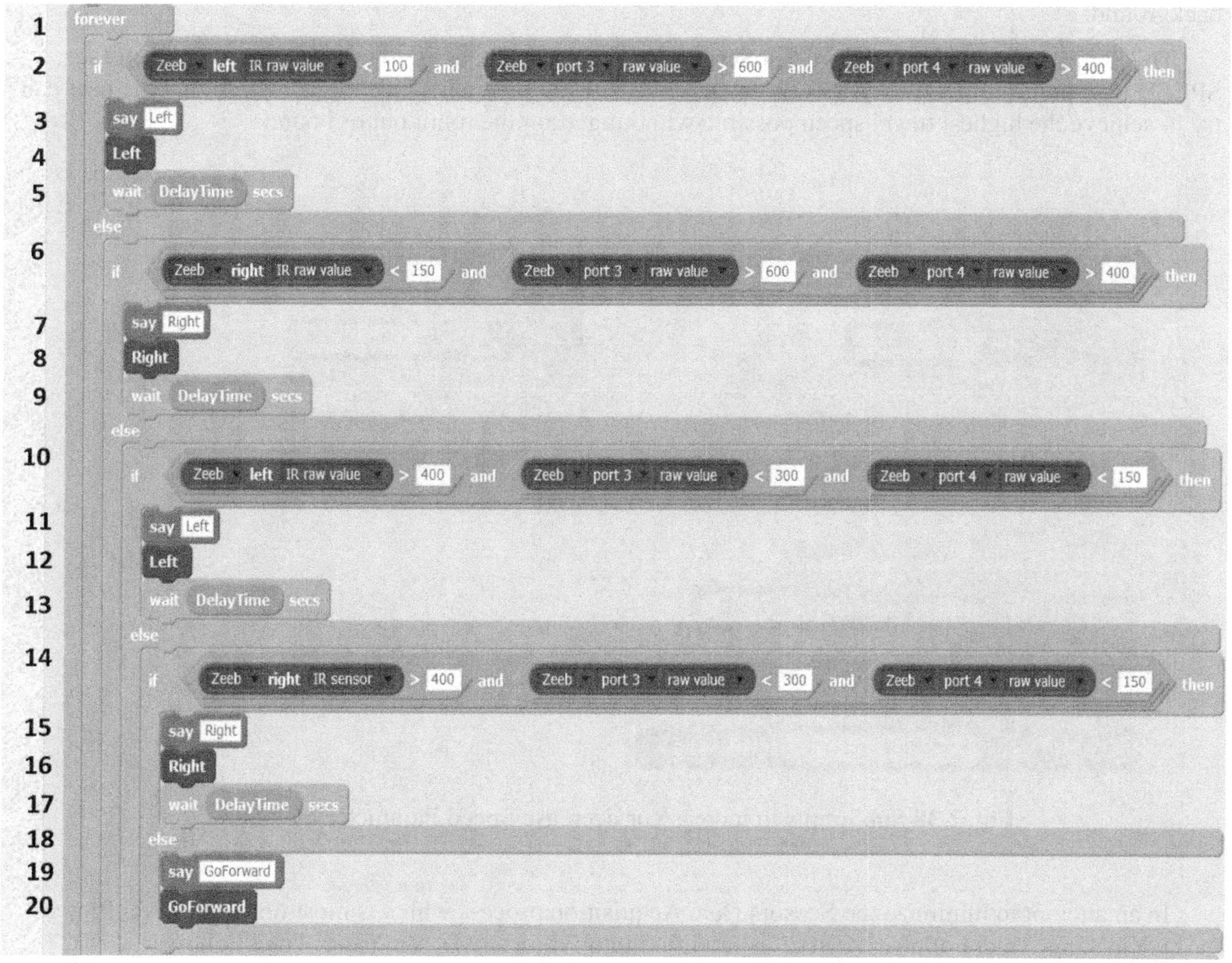

Fig. 2.37 Yin-Yang Track Following Algorithm as SCRATCH code in "AD_YinYang_TrackFollower_1.SB2".

1. Lines 2-5 correspond to the First Condition/Action pair (i.e. to turn Left into black line).
2. Lines 6-9 correspond to the Second Condition/Action pair (i.e. to turn Right into black line).
3. Lines 10-13 correspond to the Third Condition/Action pair (i.e. to turn Left into white line).
4. Lines 14-17 correspond to the Fourth Condition/Action pair (i.e. to turn Right into white line).

5. Lines 18-20 correspond to the Default Action of Going Forward.

In Project "AD_YinYang_TrackFollower_1.SB2", the DelayTime was set to zero, and the reader is encouraged to tune this value to get the best performance out of his or her robot. For the author's robot, Parameter Speed was set to 30, but the reader is also encouraged to try to get Speed as high as possible for his or her own track. In general, higher speeds are achievable for slightly curved tracks while slower speeds are needed for tracks with sharper turns.

The reader ought to note that all IR Sensors are slightly different from each other, thus each of them uses a different threshold value to detect either the white or black line, as well as the white or black background.

This project also implements two SubScripts (i.e. Functions) that use the UP-ARROW to increase SPEED by 2 units and DOWN-ARROW to decrease SPEED by 2 units (see Fig. 2.38), so that the user can try to achieve the highest travel speed possible without getting the robot out of bounds.

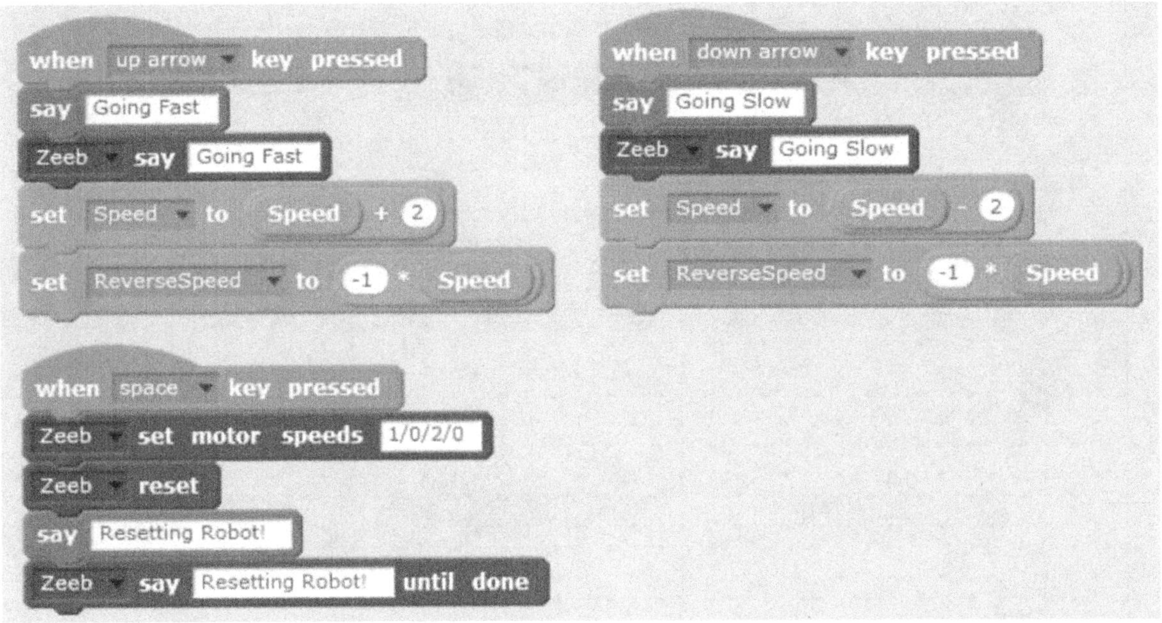

Fig. 2.38 SubScripts to increase or decrease Speed Parameter by 2 units.

In an attempt to minimize the Sensors Data Acquisition process which is most time intensive, Project "AD_YinYang_TrackFollower_1.SB2" is modified into Project "AD_YinYang_TrackFollower_2.SB2" where the Sensors Data Acquisition is done once per loop (see Fig. 2.39).

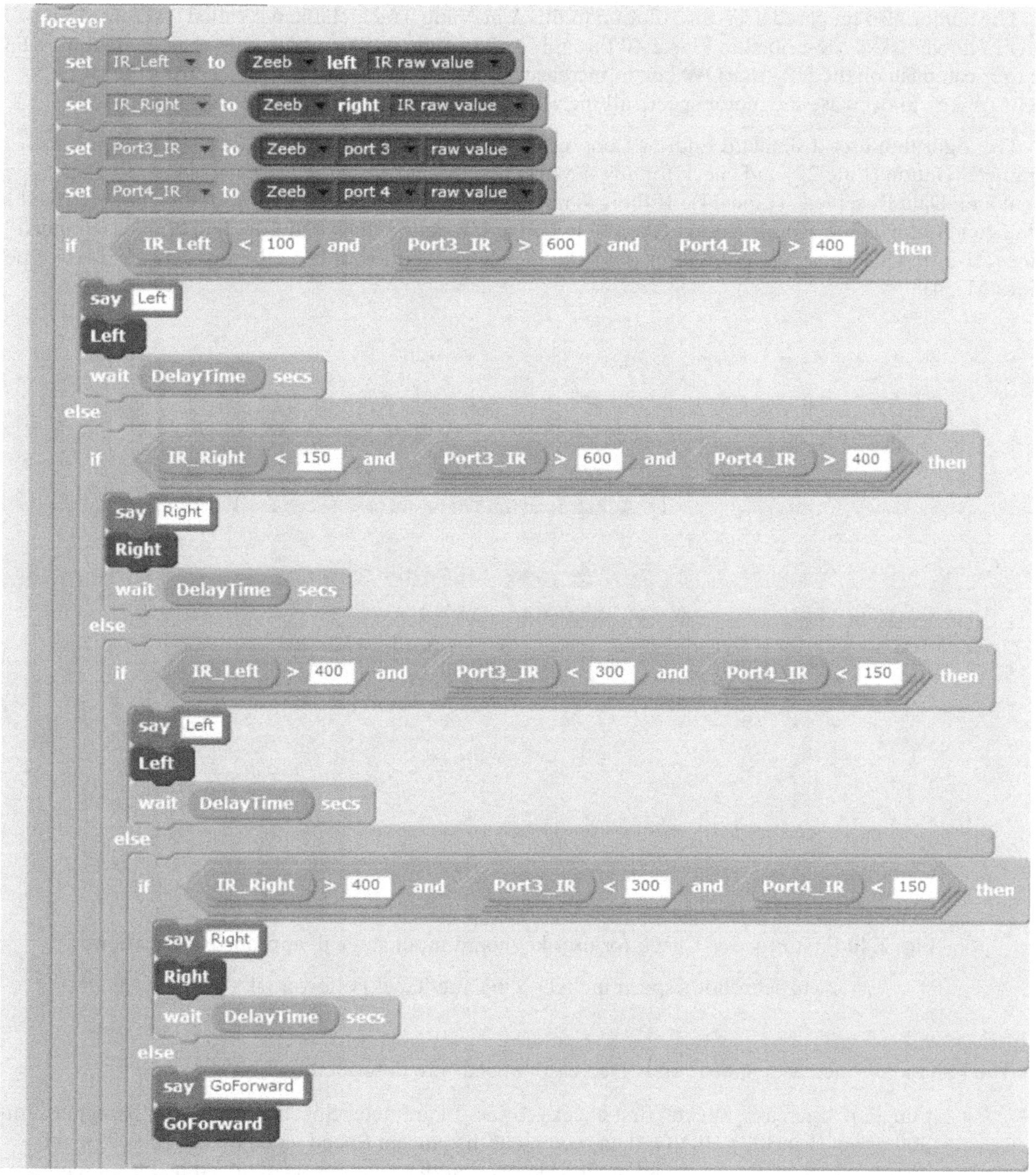

Fig. 2.39 Modified Yin-Yang Track Following Algorithm in “AD_YinYang_TrackFollower_2.SB2”.

Video 2.12 shows the runtime performances of both solutions which were about the same.

The author also prepared a TASK solution to this Yin-Yang Track challenge called "AD-YinYang-TrackFollower.tskx" (described in Figs 2.40 through 2.42). This TASK solution also has the feature where the user can push on the UP-ARROW key to increase the motor speed and conversely uses the DOWN-ARROW key to decrease the motor speed, all the while the robot is traversing the Yin-Yang track.

The algorithm uses a standard Endless Loop where the first process (see Fig. 2.40) is to clear out Parameter Button (Line 22) and check for any new keyboard input from the user, i.e. check the flag "Remocon Data Received" (Line 24). If there is new keyboard input, then save the actual data received into Parameter Button (Line 26). If Button is found to be the UP button, then call Function "Increase_Speed" (Lines 27-30). Otherwise, if Button turns out to be the DOWN button, then call Function "Decrease_Speed" (Lines 31-34).

```
20  ENDLESS LOOP
21  {
22      Button = --
23      Print Screen with Line = Speed
24      IF ( Remocon Data Received == TRUE (1) )
25      {
26          Button = Remocon RXD
27          IF ( Button == U )
28          {
29              CALL Increase_Speed
30          }
31          ELSE IF ( Button == D )
32          {
33              CALL Decrease_Speed
34          }
35      }
```

Fig. 2.40 First Process: Check for user keyboard input and call appropriate Functions to set robot's speed in "AD-YinYang-TrackFollower.tskx".

The Functions "Increase_Speed" and "Decrease_Speed" are described in Fig. 2.41:

- Function "Increase_Speed" first checks to see if Parameter Speed has reached the upper limit of 900 or not (Line 97). If NO, then increment the current Speed value by 10 (Line 99), and buzzes a La# for 0.1 sec to let the user know that the robot's speed indeed had been increased (Lines 100-102). If YES, then exit Function "Increase_Speed" with no audio feedback (i.e. "silence" is used to let the user know that Speed has reached its upper limit of 900).
- Similarly, Function "Decrease_Speed" first checks to see if Parameter Speed has reached the lower limit of 120 or not (Line 108). If NO, then decrement the current Speed value by 10 (Line 110) and buzzes a Do# for 0.1 sec to let the user know that the robot's speed had been

successfully decreased (Lines 111-113). If YES, then exit Function "Decrease_Speed" with no audio feedback to let the user know that Speed has reached its lower limit of 120.

```
95  FUNCTION Increase_Speed
96  {
97      IF ( Speed < 900 )
98      {
99          Speed = Speed + 10
100         Buzzer Timer = 0.1sec
101         Buzzer Index = La# (1)
102         WAIT WHILE ( Buzzer Timer > 0.0sec )
103     }
104 }
105
106 FUNCTION Decrease_Speed
107 {
108     IF ( Speed > 120 )
109     {
110         Speed = Speed - 10
111         Buzzer Timer = 0.1sec
112         Buzzer Index = Do# (4)
113         WAIT WHILE ( Buzzer Timer > 0.0sec )
114     }
115 }
```

Fig. 2.41 Functions "Increase_Speed" and Decrease_Speed" in "AD-YinYang-TrackFollower.tskx".

Once the process for checking the user's keyboard input is done, the tracking process starts (see Fig. 2.42). It uses an IF-ELSE-IF structure to implement 4 mutually exclusive condition/action pairs and 1 default action:

1. If the Left IR Sensor sees Black Track and Ports 3 & 4 IR Sensors see White Background, then call Function LeftTurn (Lines 37-41).
2. Else If the Left IR Sensor sees White Track and Ports 3 & 4 IR Sensors see Black Background, then call Function LeftTurn (Lines 42-46).
3. Else If the Right IR Sensor sees Black Track and Ports 3 & 4 IR Sensors see White Background, then call Function RightTurn (Lines 47-51).
4. Else If the Right IR Sensor sees White Track and Ports 3 & 4 IR Sensors see Black Background, then call Function RightTurn (Lines 52-56).
5. Else the robot needs to go Forward – Lines 58-61. The robot also needs to start its trek straddling on the black/white line.

```
// Algorithm to traverse Yin-Yang Track
IF ( IR Left <= 100 && PORT[3]:IR Sensor >= 600 && PORT[4]:IR Sensor >= 400 )
{
    // Turns left when ONLY the left sensor detects black track & NIR_3+NIR_4 detect white background
    CALL LeftTurn
}
ELSE IF ( IR Left >= 400 && PORT[3]:IR Sensor <= 300 && PORT[4]:IR Sensor <= 150 )
{
    // Turns left when ONLY the left sensor detects white track & NIR_3+NIR_4 detect black background
    CALL LeftTurn
}
ELSE IF ( IR Right <= 150 && PORT[3]:IR Sensor >= 600 && PORT[4]:IR Sensor >= 400 )
{
    // Turns right when ONLY the right sensor detects black track & NIR_3+NIR_4 detects white background
    CALL RightTurn
}
ELSE IF ( IR Right >= 400 && PORT[3]:IR Sensor <= 300 && PORT[4]:IR Sensor <= 150 )
{
    // Turns right when ONLY the right sensor detects white track & NIR_3+NIR_4 detects black background
    CALL RightTurn
}
ELSE
{
    // Moves forward for all other cases.
    CALL Forward
}
```

Fig. 2.42 IF-ELSE-IF structure to allow robot to traverse Yin-Yang track in "AD-YinYang-TrackFollower.tskx".

As expected, Video 2.13 shows that this TASK solution outperforms the previous SCRATCH solutions. For completeness sake, a R+SCRATCH version of this project is also included in the sample codes for this book - "AD-YinYang-TrackFollower_R.SB2". As expected, it performs less well than the TASK version and about the same as the Edbot's version.

Thus, the use of the Edbot tool for autonomous and sensor-intensive SCRATCH projects requires some algorithm analysis and fine adjustments to get the best performance out of your robot, although still behind the TASK approach. However, TASK cannot control multiple robots in the same piece of code as Edbot can, as shown in the next section 2.4.7.

2.4.7 Zeeb-Zob Dowel Search Game

This section assumes that the user has 2 DREAM robots, namely Zeeb and Zob for the author (see Fig. 2.43). Zeeb is essentially the Avoider with 2 added NIR Sensors IRSS-10 on Ports 3 and 4, and Zob is the Buffalo with 1 added NIR Sensor IRSS-10 on Port 3.

Fig. 2.43 DREAM Robots Zeeb (left) and Zob (right).

The goal of this project is to show how easy it is for Edbot to control two robots concurrently and for SCRATCH to share "control events" between Sprites. The "AD_ZeebZob_DowelSearch.SB2" Project uses 3 Sprites as shown in Fig. 2.44: "Zeeb" and "Zob" represent the respective robots, while the "Cat" represents the Game Controller.

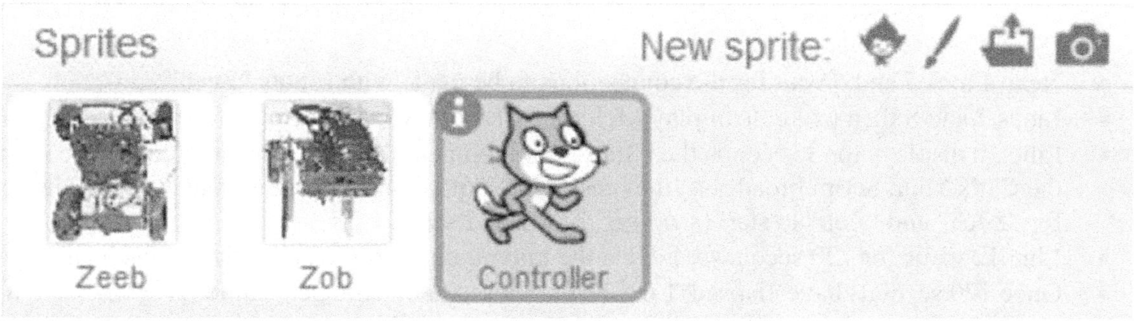

Fig. 2.44 Project "AD_ZeebZob_DowelSearch.SB2" uses 3 Sprites: Zeeb, Zob & Cat Controller.

The Game Controller "Cat" has 2 Scripts as shown in Figs. 2.45 and 2.46. The "Start" of the game is controlled by the Cat's Main Script as shown in Fig. 2.45 whereas:

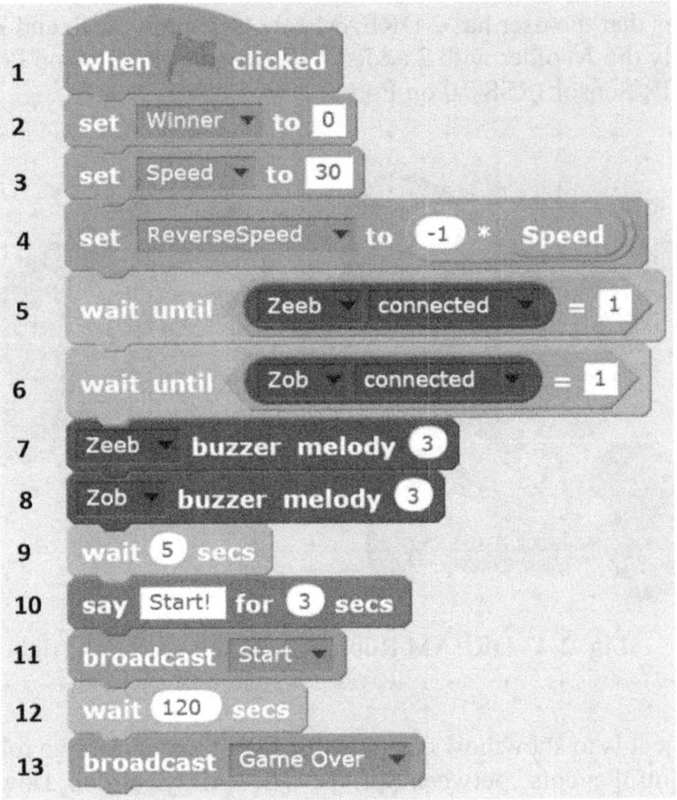

Fig. 2.45 Main Script for Game Controller "Cat" in "AD_ZeebZob_DowelSearch.SB2".

- When the Flag is clicked (Line 1), Lines 2-4 set the initial values to Variables "Winner", "Speed" and "ReverseSpeed".
- Next, Lines 5 and 6 wait for the connections to be made with robots "Zeeb" and "Zob".
- Lines 7 and 8 then make them play Melody 3. Line 9 waits 5 seconds for the melodies to finish.
- Line 10 displays for 3 seconds the "Start!" balloon near the "Cat" Sprite in the Stage Area, then the Cat's Main Script broadcasts the message "Start" (Line 11) which makes the Main Scripts for "Zeeb" and "Zob" to start (see Figs. 2.47 and 2.49).
- Line 12 waits for 120 seconds which is the maximum time span allowed for each game session.
- Once 120 seconds have elapsed. Line 13 broadcasts the message "Game Over" to wrap up the current game session (see Fig. 2.46).

The Cat's Game Ending thread is described in Fig. 2.46 whereas:

- Line 14 picks up the message "Game Over" which can be sent by 3 Scripts: Cat's Main Script (Fig. 2.45), Zeeb's Main Script (Fig. 2.47) and Zob's Main Script (Fig. 2.49).
- Lines 15-16 make Zeeb and Zob play Melody 4.

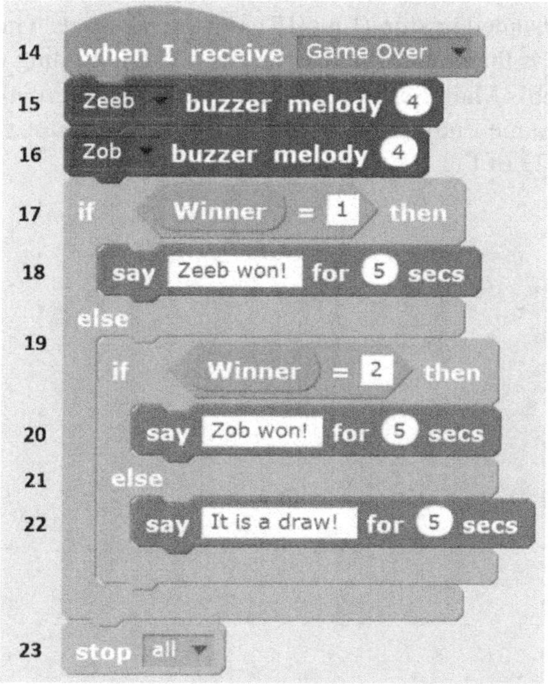

Fig. 2.46 How Game Controller "Cat" ends a game session in project "AD_ZeebZob_DowelSearch.SB2".

- The IF-ELSE-IF structure (Lines 17-22) uses the current value of Variable "Winner" to determine if either Zeeb (Lines 17-18) or Zob (Lines 19-20) has won, or that it is a draw (Lines 21-22)?
- Line 23 then stops all threads to end the current game session.

Fig. 2.47 illustrates the Main Script for the Zeeb Sprite:

- Line 1 starts when this thread receives the message "Start" (from the Cat Controller).
- Lines 2-4 initialize the values for Variables "Dowel_Distance_L", "Dowel_Distance_R" and "Start_Direction" (which is a random value, either "1" or "2").
- An IF-ELSE structure (Lines 5-7) makes Zeeb spin Left (Line 6) or Right (Line 7) depending on the current value of "Start_Direction". Please see Fig. 2.48 for the definition of Zeeb's User Blocks (Zeeb_Left, Zeeb_Right and Zeeb_Forward).
- Next a FOREVER Loop (Lines 8-19) is entered, where essentially the latest IR values from Ports 3 and 4 are saved into their respective Variables Port3_IR (Line 9) and Port4_IR (Line 10), every time that this loop is iterated.
- The more complex IF structure (Lines 11-19) first checks if the physical dowel is found by either Port 3 or Port 4 IR sensors (Line 11). If this condition is FALSE, then this thread loops back to Line 9. If this condition is TRUE, then Zeeb goes forward for 2 seconds (Lines 12-13).
- After Line 13 finishes its 2 seconds wait time, Lines 14-16 determine if Zeeb is the true winner, i.e. when Variable Winner is still set to zero (Lines 14-15), or if it is a draw (Line 16), i.e. when Zob also manages to get to a similar result (Winner = 2) at the same point in time. Please note that there are two threads (two Forever Loops to be exact – Figs. 2.47 and 2.49) running in parallel during this time span.

- Next Zeeb is commanded to stop (Line 17) and the message "Game Over" is broadcasted (Line 18). This message is then picked up by the Cat's Game Ending thread – Line 14 of Fig. 2.46).
- Line 19 makes Zeeb's Main Script wait for 10 seconds, essentially Zeeb does nothing while waiting the Cat's Game Ending thread to finish its job and stop ALL processes including Zeeb's Main Script (Line 23 of Fig. 2.46).

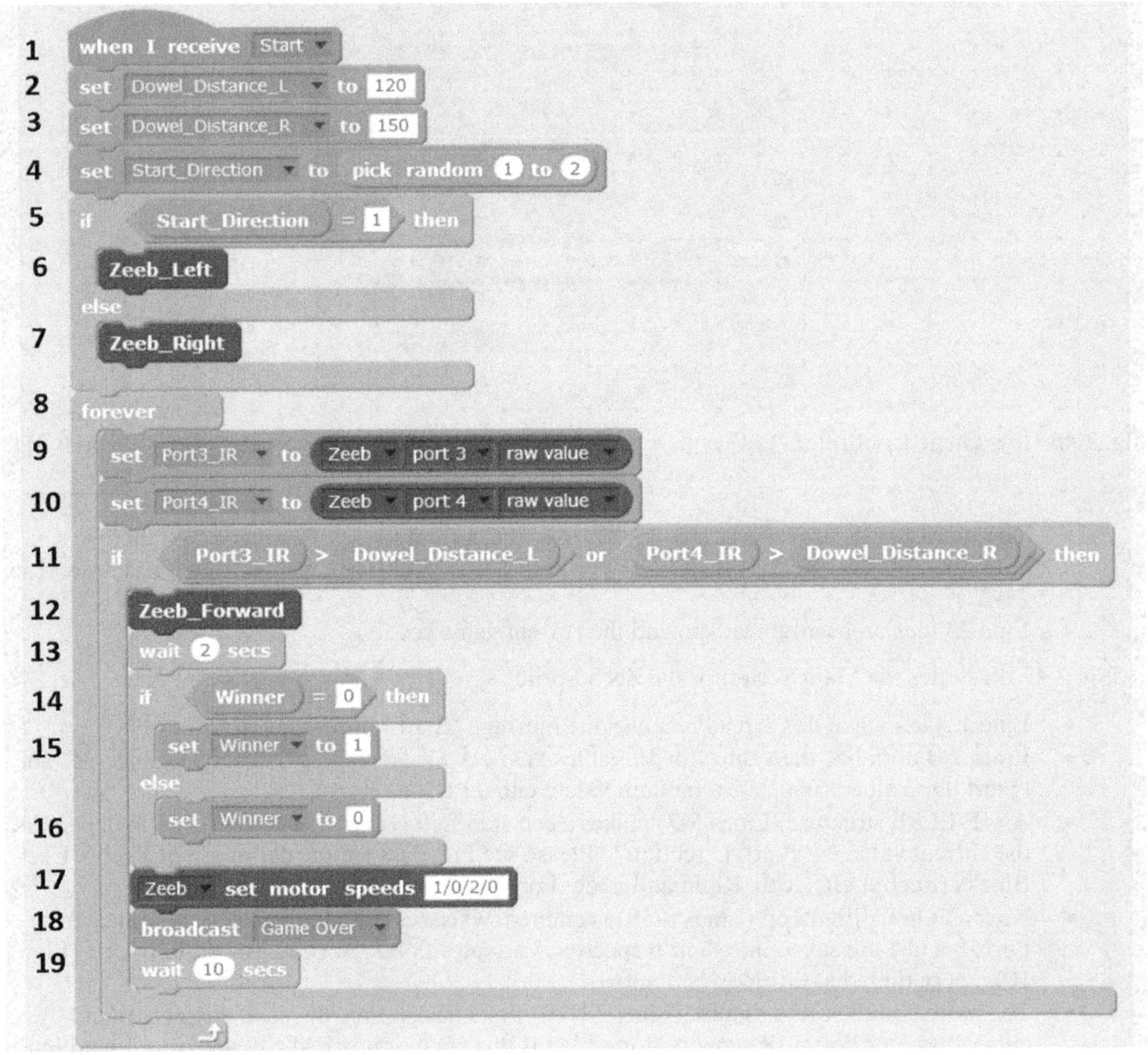

Fig. 2.47 Main Script for Zeeb in "AD_ZeebZob_DowelSearch.SB2".

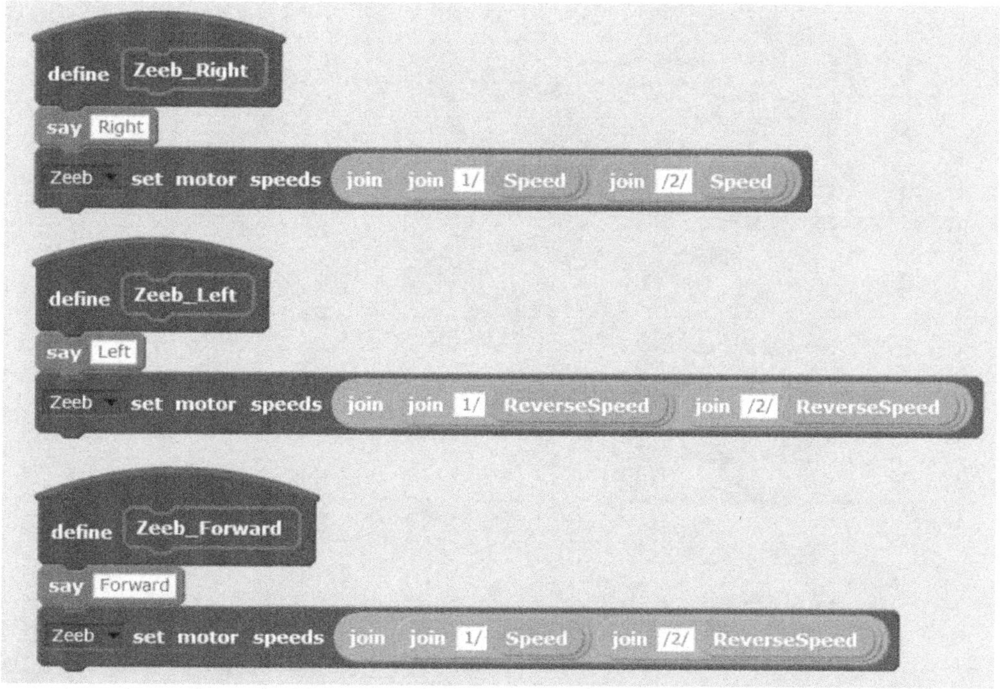

Fig. 2.48 User-Defined Blocks for Zeeb in "AD_ZeebZob_DowelSearch.SB2".

Fig. 2.49 illustrates the Main Script for the Zob Sprite:

- Line 1 starts when this thread receives the message "Start" (from the Cat Controller).
- Lines 2-3 initialize the values for Variables "Dowel_Distance_C" and "Start_Direction" (which is a random value, either "1" or "2").
- An IF-ELSE structure (Lines 4-6) makes Zob spin Left (Line 5) or Right (Line 6) depending on the current value of "Start_Direction". Please see Fig. 2.50 for the definition of Zob's User Blocks (Zob_Left, Zob_Right and Zob_Forward).
- Next a FOREVER Loop (Lines 7-17) is entered, where essentially the latest IR value from Port 3 is saved into Variable Port3_IR (Line 8), every time that this loop is iterated.
- The more complex IF structure (Lines 9-17) first checks if the physical dowel is found by the Port 3 IR sensor (Line 9). If this condition is FALSE, then this thread loops back to Line 8. If this condition is TRUE, then Zob goes forward for 2 seconds (Lines 10-11).
- Then Lines 12-14 determine if Zob is the true winner (Lines 12-13) or if it is a draw (Line 14).
- Next Zob is commanded to stop (Line 15) and the message "Game Over" is broadcasted (Line 16) and then picked up by the Cat's Game Ending thread (Line 14 of Fig. 2.46).
- Line 17 makes Zob's Main Script waits for 10 seconds, essentially Zob does nothing while waiting the Cat's Game Ending thread to finish its job and stop ALL processes, including Zob's Main Script (Line 23 of Fig. 2.46).

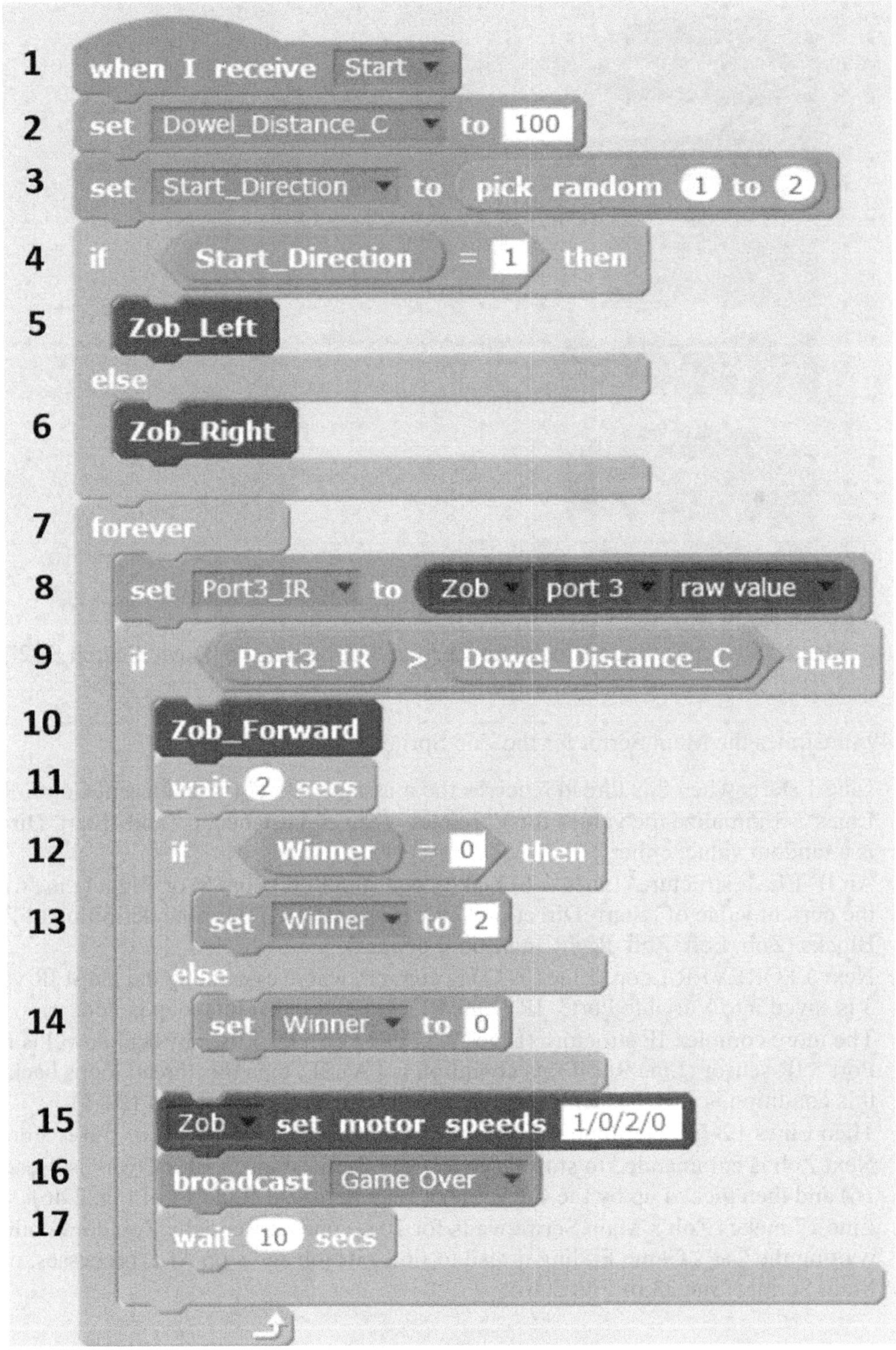

Fig. 2.49 Main Script for Zob in "AD_ZeebZob_DowelSearch.SB2".

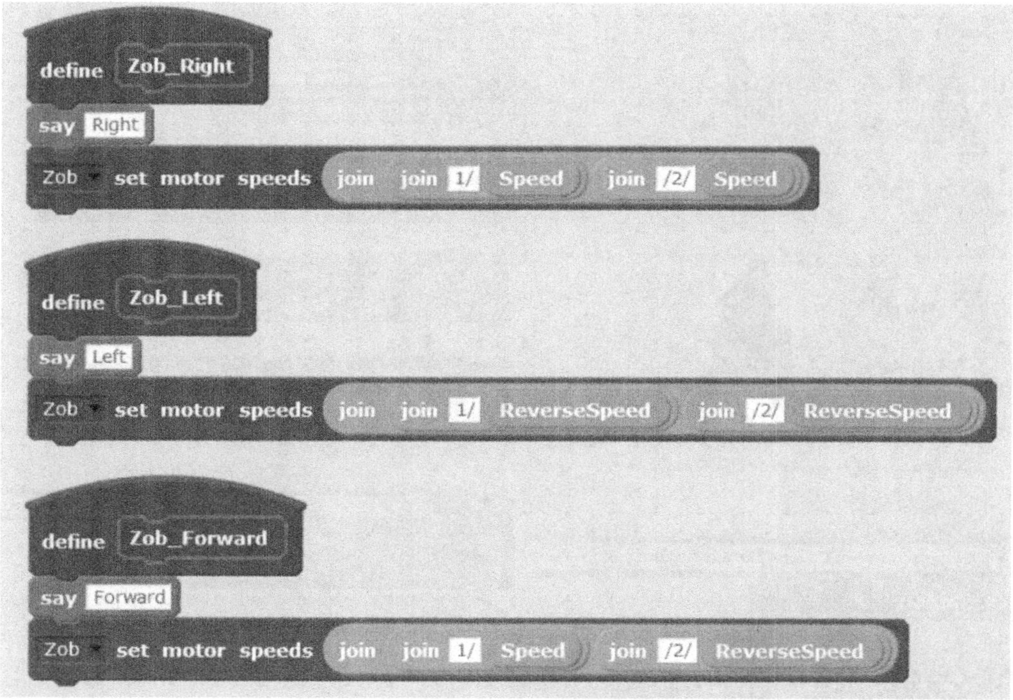

Fig. 2.50 User-Defined Blocks for Zob in "AD_ZeebZob_DowelSearch.SB2".

Video 2.14 illustrates the run-time performance of this project with Zeeb and Zob.

2.4.8 Mixed Multiple-Robots Control with Edbot

Although the author suspects that not many readers have much ROBOTIS hardware at their disposal as the author has, but all readers need to know about this unique feature of the Edbot tool. At this point in time, Edbot is the **only** software tool that allows the control of multiple and mixed robots (DREAM and/or MINI) together in a single SCRATCH 2 project such as "AD-FamilyZZZZ.SB2" (see Fig. 2.51).

The reader can see from Fig. 2.51 that it is quite straight forward to control multiple and mixed-type robots by just laying the appropriate "named" Action blocks that belong to each type of robots (MINI or DREAM) following a user-defined action script and let the Edbot tool performs its "magic" at run time. Zira and Zork are MINI robots while Zeeb and Zob are DREAM robots in this project.

Please watch this YouTube video https://www.youtube.com/watch?v=yUpx42eNR2I for a performance of the "AD-FamilyZZZZ.SB2" project.

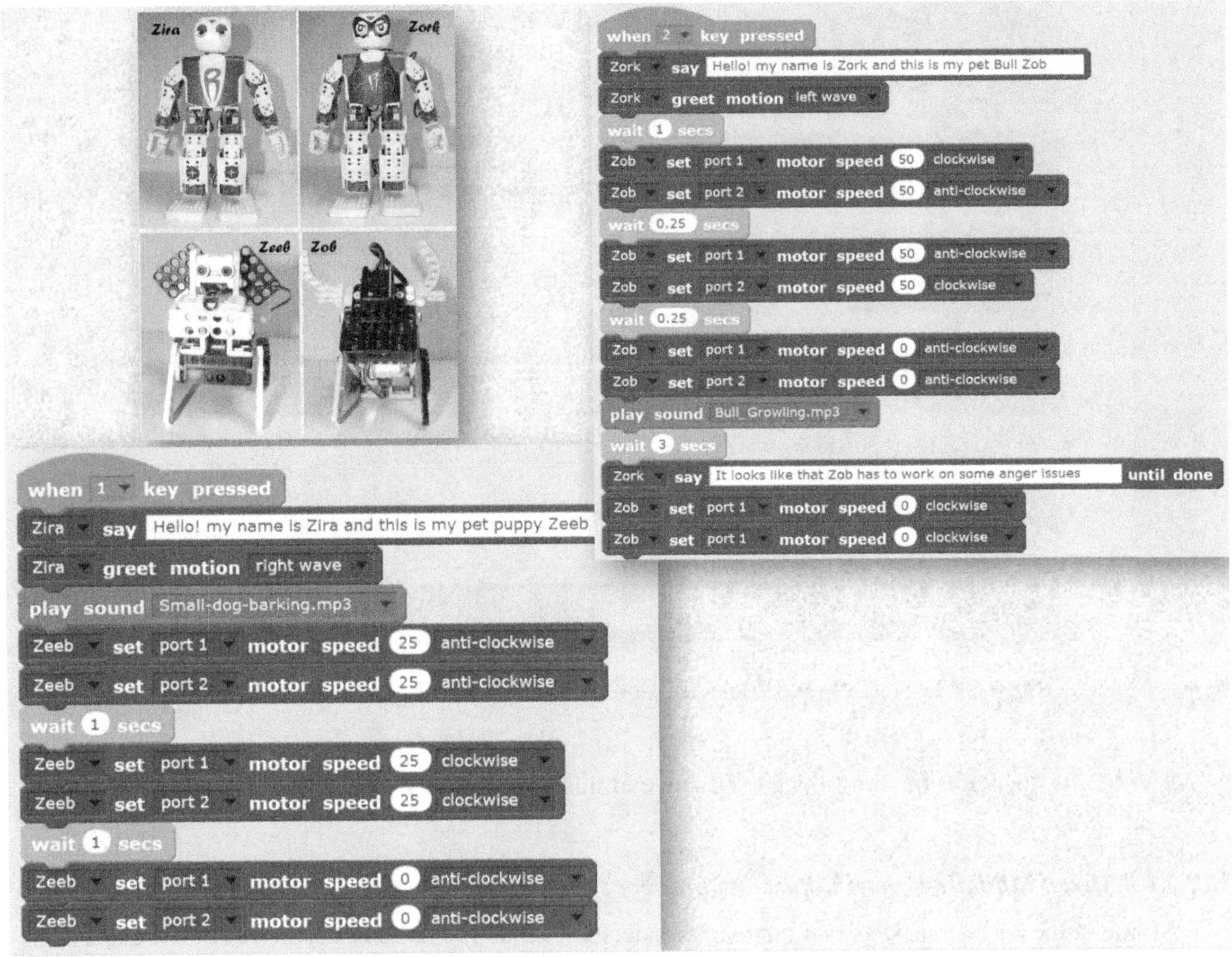

Fig. 2.51 Multiple Robots Control with SCRATCH and Edbot.

In short, Edbot is such a unique and powerful tool for controlling multiple ROBOTIS MINI and DREAM robots when sensor-intensive autonomous behaviors are not (much) required in one's SCRATCH 2/3 project. Once the Edbot folks resolve the data packet flow issue, the author expects that Edbot would remain in the top spot of the Educational Robotics software arena for quite some time.

2.4.9 Using Servo Motor SM-10 in Video Game

This project is an adaptation from a SCRATCH example written by ROBOTIS for the IoT system (http://www.robotis.com/model/page.php?co_id=prd_iot). Currently (Spring 2019), the IoT kit is only available in South Korea. It simulates a Car racing on a Track which is being scrolled up in the Stage Area of the SCRATCH IDE (see Fig. 2.52). The user can steer the Car by using a Servo Motor (SM-10) as a Steering Column (i.e. as a rotational position encoder) (see Fig. 2.53).

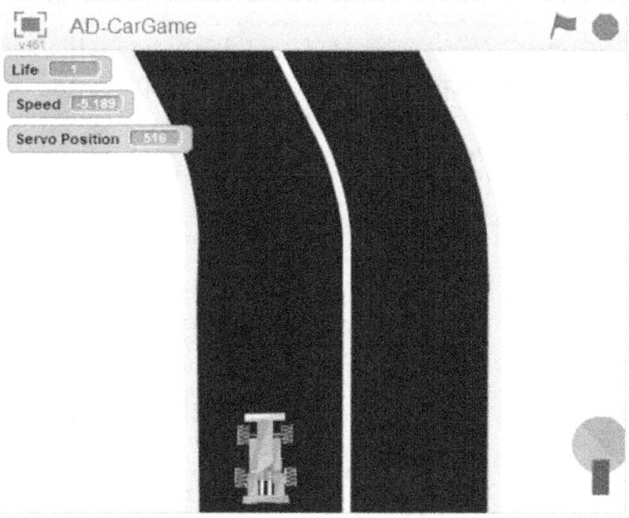

Fig. 2.52 Stage Area of SCRATCH IDE for Project "AD-CarGame.SB2".

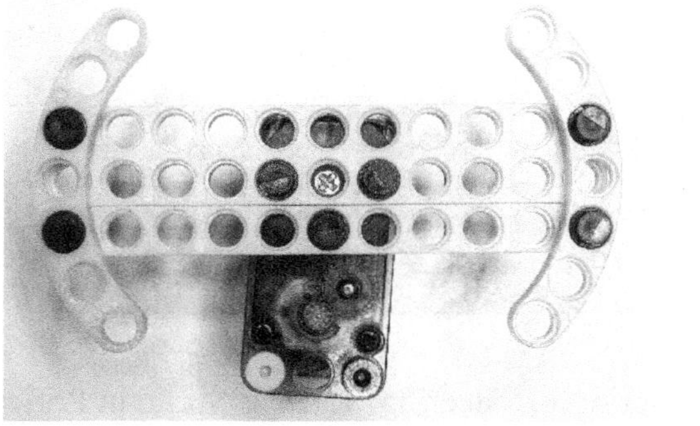

Fig. 2.53 Servo Motor SM-10 used as a "Steering Column" for Project "AD-CarGame.SB2".

Let's first look at how to use the SM-10 to "steer" the Car Sprite with Project "AD-Car.SB2" (see Fig. 2.54), with the Car Sprite being a graphics image with an original size of 233 x 154 pixels which is shrunk down to about 70 x 48 pixels when displayed on the Stage Area which has the standard size of 480 x 360 pixels (see Fig. 2.52):

- Line 2 puts the Car Sprite in the foreground of the Stage Area.
- Line 3 points the front of the Car upwards.
- Lines 4 and 5 set both X & Y positions of the Car to 0, i.e. put it in the center of the Stage Area.
- Line 6 sends the servo's horn into its 150-degree position (i.e. straight up – see web link http://support.ed.bot/edbot-dream-scratch2.html).
- Line 7 waits for 2 seconds for the horn to reach the 150-degree position from wherever it was at start.
- Line 8 turns the Port 3 servo off so that the user can rotate the steering column by hand, and Line 9 announces to the user that the Car is ready to be steered.
- Lines 10-13 is an Endless Loop where it keeps the car at position (X_Pos_Car, Y_Pos_Car) in the Stage Area (Line 11), while the Variables X_Pos_Car and Y_Pos_Car are being updated by

4 independent SubScripts described in Fig. 2.55. At this point in time, the user is expected to manipulate the Steering Column as needed, and this action will change the Horn Position of the Servo Motor (i.e. Variable "Servo Position") which is updated by Line 12 as it reads in the latest Servo Position in "raw mode" (values between 0 and 1023) and saves this value in "Servo Position".

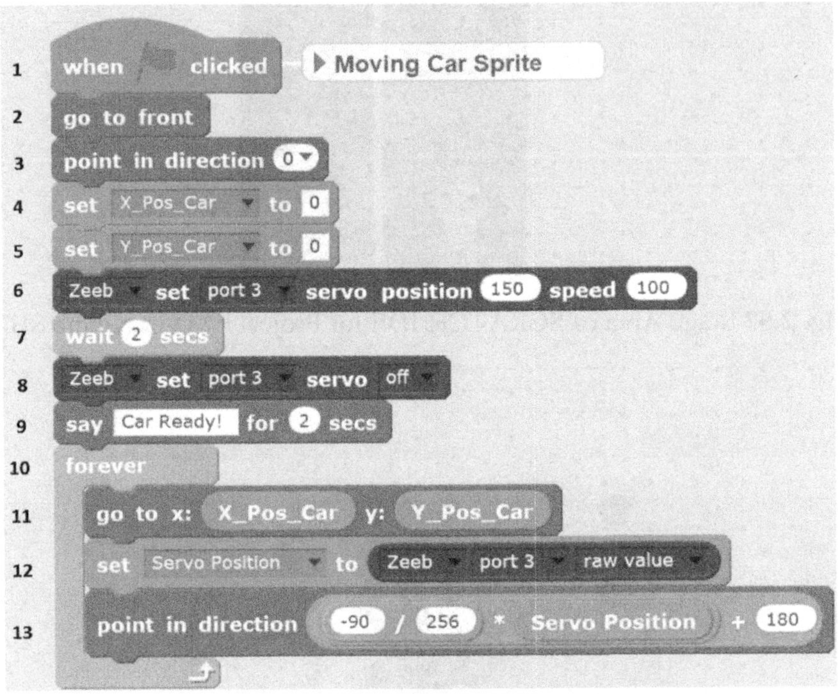

Fig. 2.54 Main Script for "steering" the Car Sprite in Project "AD-Car.SB2".

- Line 13 converts this "raw value" into a directional angle (in degrees) that SCRATCH would be expecting with the algebraic formula: **(-90/256) * (Servo Position) + 180**. For example, when the Servo was sent to the Starting Position 150 (Edbot's convention), the corresponding Raw Value would read in as "512". This formula now becomes this expression (-90/256)*(512) + 180, which evaluates to 0 and this value is fed into the "Point In Direction" Block which then positions the Car Sprite's Front End straight up in the vertical direction (i.e. along the Y-axis). If the user rotates the Steering Column to the physical left of the "512" position, "Servo Position" would have a value larger than 512 which makes the above formula produce a negative value. Therefore, the Car Sprite's Directional Angle is now a negative value, and the Car Sprite will be rotated to the left of the vertical line (Y-axis) in the Stage Area. Vice-versa, if the user turns the Servo to the right of the "512" position, "Servo Position" would be a number smaller than 512, and the formula yields a positive value for Directional Angle. Therefore, the Car Sprite will be rotated to the right of the Y-axis.

Fig. 2.55 shows 4 independent SubScripts which are activated by the 4 ARROW keys:

- Each time that the "UP ARROW" is pushed, "Y_Pos_Car" is incremented by 2, i.e. the Car Sprite moves upwards in the Stage Area.
- Each time that the "DOWN ARROW" is pushed, "Y_Pos_Car" is decremented by 2, i.e. the Car Sprite moves downwards in the Stage Area.
- Each time that the "RIGHT ARROW" is pushed, "X_Pos_Car" is incremented by 2, i.e. the Car Sprite moves right in the Stage Area.
- Each time that the "LEFT ARROW" is pushed, "X_Pos_Car" is decremented by 2, i.e. the Car Sprite moves left in the Stage Area.

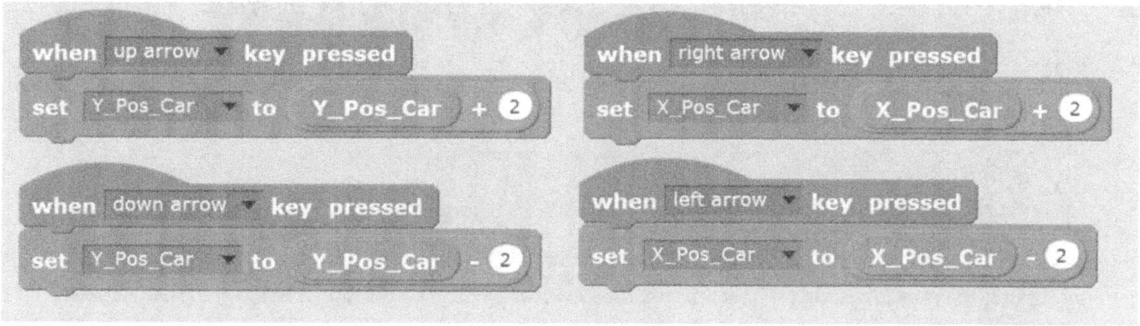

Fig. 2.55 4 SubScripts for moving the Car Sprite U-D-L-R in Project "AD-Car.SB2".

Lastly, Fig. 2.56 shows how the "Space" bar would first reset the Servo Motor to the "150" position (Lines 16-17) and waits 2 seconds for it to get there (Line 18). Next the Raw Servo Position is read in (Line 19) and is used later in Line 23 to align the Car Sprite along the Y-axis. Lines 20-22 also help by putting the Car at the center of the Stage Area. Line 24 stops all running scripts (including the Main one which was still running the Forever Loop).

Video 2.15 demonstrates how this Car Steering Project performs in real-time.

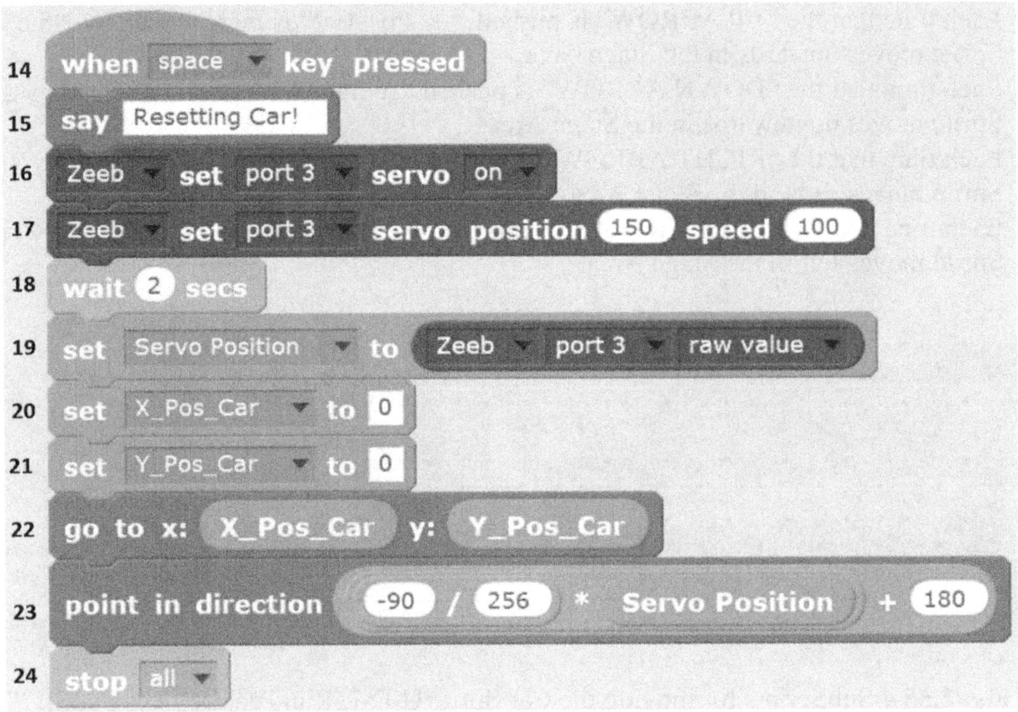

Fig. 2.56 SubScript showing how the "Space" bar resets the servo motor and the Car Sprite positions in Project "AD-Car.SB2".

Next, let's examine how the Track Sprite (940 x 5057 pixels) is manipulated to give the user the impression that the Car Sprite is travelling on the track (see the previous Fig. 2.52). Please note that the Track Sprite is larger in pixel size than the Pixel Space reserved to the Stage Area which is only 480 x 360 pixels.

The general approach is to show the bottom part of the Track within the Stage Area at start, and then to scroll the Track "downward" so that the user got the impression that the Car is travelling "upward" the Track. Another needed feature is how to use the real-time Servo Position to adjust the sideways drift of the Track to accommodate for "turning maneuvers", i.e. when the Track displays "curves" in the Stage Area.

Fig. 2.57 displays the Main Script for Project "AD-Track.SB2":

- Lines 2-3 send the Servo on Port 3 to its 150 position, i.e. reset the Steering Column to its starting position.
- Line 4 orients the Track Sprite so that its Starting Line is at the bottom of the Track Sprite and its Finish Line is at the top of the Track Sprite.

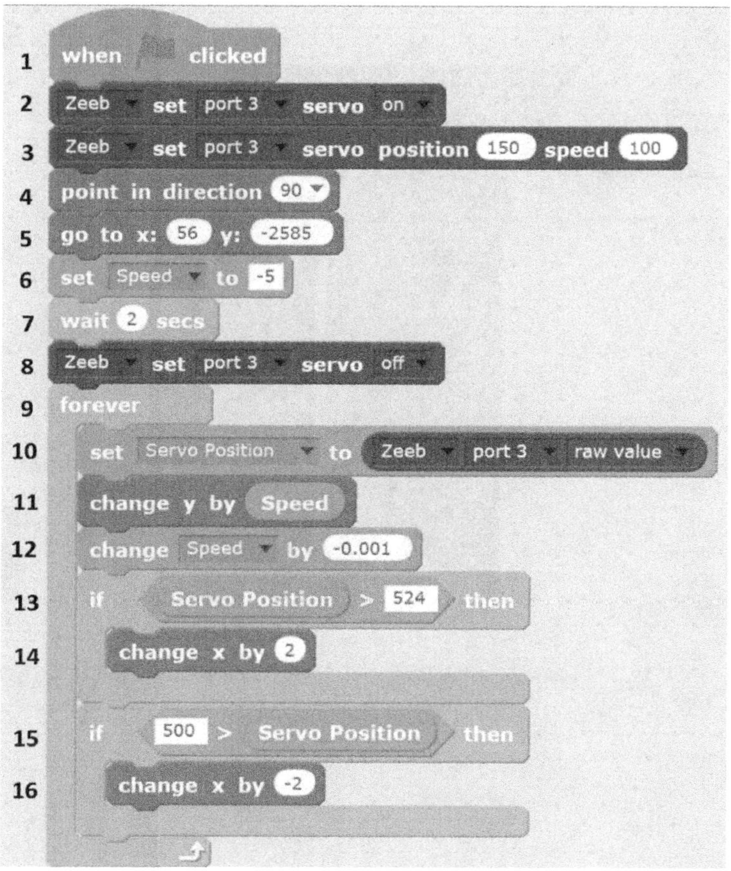

Fig. 2.57 Main Script for "scrolling" the Track Sprite in Project "AD-Track.SB2".

- Line 5 sets the bottom edge of the Track Sprite and the "actual" graphical track within the viewport of the Stage Area (see Fig. 2.58). Thus, to simulate the car progressing from the Start Line to the Finish Line, we will have to make the Y coordinate get more and more negative.
- Line 6 sets Variable Speed to the value "-5".
- Line 7 waits for 2 seconds for the Servo Motor to reach its 150 position as commanded previously by Line 3 and Line 8 turns the Servo Off so that the user can rotate it manually.
- The Forever Loop (Lines 9-16) does the "scrolling down" and "sideways drift" of the Track Sprite to simulate the Car's progress from Start to Finish.
- Line 10 set Variable "Servo Position" to the Raw Value [0-1023] read in from Port 3's Servo Motor (which should be around "512" at start).
- Line 11 does the actual scrolling down by a certain number of pixels represented by the current value of Variable Speed (which starts out at "-5" – see Line 6). Line 12 shows that for each iteration of the loop, Speed gets more and more negative – this is to simulate that the Car's speed increases with the game's run time to make the game more exciting!

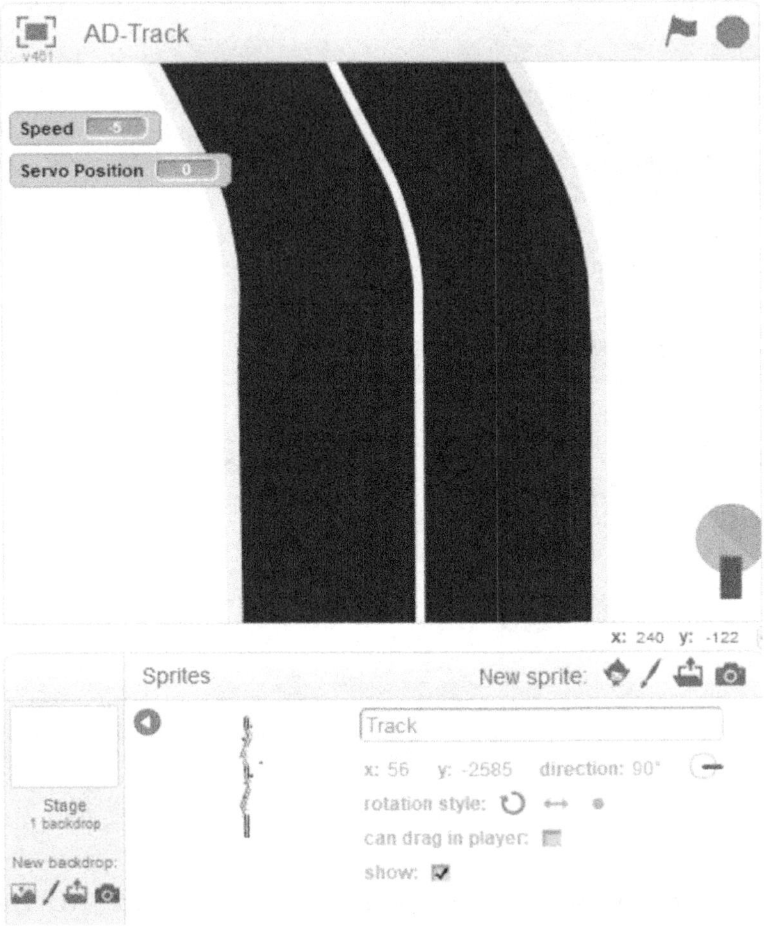

Fig. 2.58 Starting Orientation and Position for the Track Sprite in Project "AD-Track.SB2".

- Next the two parallel IF structures (Lines 13-14 and 15-16) use the current value of Variable "Servo Position" to create the Left and Right drifts of the Track Sprite to visually help the user negotiates a Left or Right curve of the Track as displayed within the Stage Area at certain moments in the game (at the same time as the Track is being scrolled down by Lines 11-12), for example:
  - When the user sees a Left curve coming up, the user would naturally rotate the Steering Column to the Left, because he/she wants the Car Sprite to rotate to the Left of its Y axis (effect not shown in this "Track" Project) to take on the curve. This action would then produce a value larger than 524 for Variable "Servo Position", triggering the IF on Line 13. Furthermore, visually, the user would then also expect the Track to drift slightly in the other direction (i.e. to the Right) which can be achieved by making the X position more positive (by Step of "+2" as shown in Line 14).
  - Conversely, when the user sees a Right curve appearing, he/she would turn the Steering Column to the Right, therefore "Servo Position" would decrease its value to less than "500", triggering the IF condition on Line 15 and then Line 16 would drift the Track to the Left by decreasing the X position by Step of "-2".

Fig. 2.59 displays the Reset feature triggered by the user pressing on the Space Bar whereas:

- Line 18 stops the scrolling process by setting Speed to zero.
- Line 19 sets the Track bottom into view of the Stage Area.
- Lines 20-23 reactivate the Servo Motor, send it to its Start Position of "150", wait for 2 seconds for it to get there and then turn this Servo OFF.
- Finally, all Scripts are stopped with Line 24.

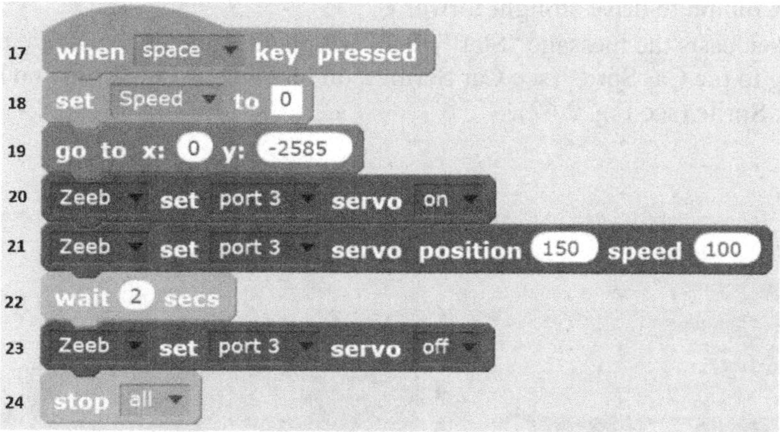

Fig. 2.59 SubScript showing how the "Space" bar resets the servo motor and the Track Sprite position in Project "AD-Track.SB2".

Video 2.16 illustrates the run-time performance of Project "AD-Track.SB2". Now, we are ready to combine the above complementary Projects to obtain a complete car game "AD-CarGame.SB2". As expected, this Project has two Sprites: Car and Track (see Fig. 2.60).

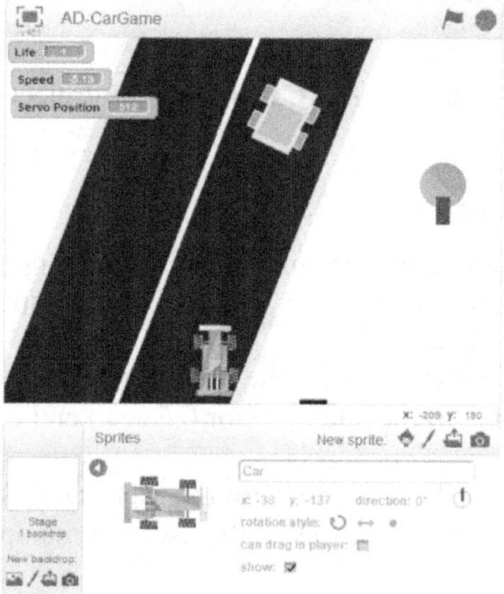

Fig. 2.60 Screenshot of Stage Area for Project "AD-CarGame.SB2".

Fig. 2.61 shows Part 1 of the two Scripts that pertain to the Car Sprite: a Main one, Car Script 1, which Broadcasts to Car Script 2 (and also the Main Script for the Track Sprite).

Car Script 1 can be triggered by the Flag Hat Block (Line 1) which leads to:

- Line 2 sets Variable "Life" to a value of 3, meaning that the user has 3 chances to finish the race successfully (i.e. to reach the Finish Line).
- Line 3 pushes the Car Sprite to the Foreground of the Stage Area.
- Lines 4 and 5 turns on Port 3 Servo Motor and sends to its "150" Position, i.e. resets the Steering Column to drive straight forward.
- Line 6 broadcasts the message "Start" to all Scripts that are set to receive this message: one belonging to the Car Sprite (see Car Script 2 on left side of Fig. 2.61) and another belonging to the Track Sprite (see Fig. 2.62).

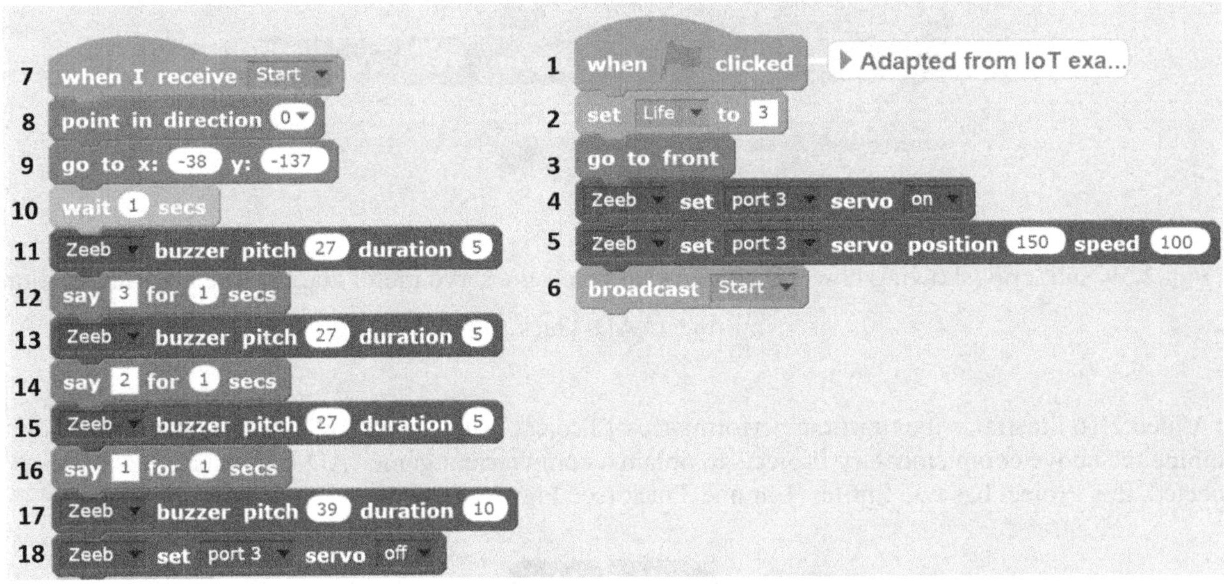

Fig. 2.61 Car Script 1 and Part 1 of Car Script 2 in Project "AD-CarGame.SB2".

Car Script 2 thus begins with Line 7 in Fig. 2.61:

- Line 7 picks up the "Start" message, and this Script rolls on to Lines 8 and 9 which set the Start Pointing Direction for the Car Sprite to "0", i.e. Upward, and its Start X-Y Positions (-38, -137) – as shown way back in Fig. 2.52.
- Line 10 waits for 1 second (mainly for the user to get situated).
- Lines 11-16 activate the Buzzer Pitch 27 (for 0.5 sec.) and coordinate with a countdown from 3 to 1.
- Line 17 plays Pitch 39 for 1 second to warn the user that this Virtual Car Race is about to begin.
- Line 18 turns Port 3's Servo OFF, so that the user can manipulate the Steering Column as needed.

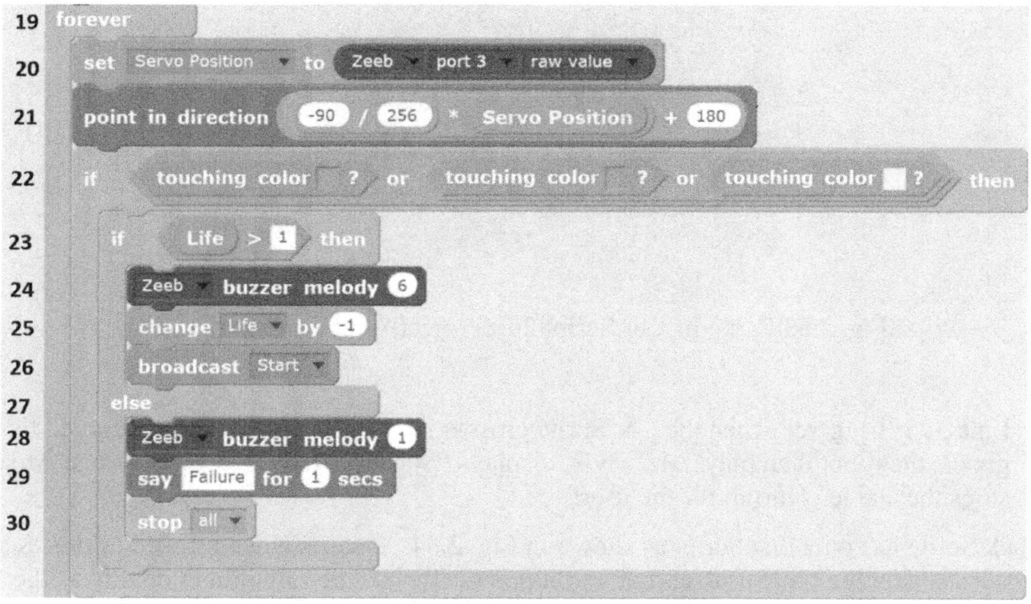

Fig. 2.62 Part 2 of Car Script 2 in Project "AD-CarGame.SB2".

Fig. 2.62 shows Part 2 of Car Script 2:

- Line 19 starts a Forever Loop which takes care of 3 tasks: 1) steering the Car, 2) checking if it bumps into any obstacle along the way, and 3) checking whether it crosses over the Finish Line (Fig. 2.63).
- Lines 20 and 21 steers the Car Sprite as commanded by the user via his/her manipulation of Port 3's Servo Motor (as explained in depth in Project "AD-Car.SB2").
- Lines 22 through 30 deal with situations when the Car Sprite bumps into 3 possible kinds of obstacles present at various locations within the Track Sprite. This task is accomplished by checking whether the Car Sprite "touches" any pixel of either color from the set of Orange, Blue or Yellow colors. "Orange" and "Blue" can represent another "Car" or "Road Sign" while "Yellow" represents the two sides of the Black road (see Fig. 2.60). These situations trigger the IF in Line 22 which in turn triggers the IF-ELSE block checking on the Current Life count for the user (Lines 23 to 30).
- If Variable Life is greater than 1 (Line 23), the robot would play Melody 6 (line 24), and Variable Life gets decremented by 1 count, and this Script broadcasts "Start" again and effectively makes Car Script 2 restart at Line 7 of Fig. 2.61 one more time.
- If Variable Life as a value of "0", the ELSE on Line 27 gets triggered and the robot plays Melody 1, displays "Failure" on the Stage Area for 1 sec. and then Line 30 stops ALL Scripts and thus effectively ends this Car Game.

However, if the IF condition of Line 22 did not realize (meaning that no obstacle had been hit by the Car Sprite so far), we enter Part 3 of Car Script 2 as shown in Fig. 2.63:

Fig. 2.63 Part 3 of Car Script 2 in Project "AD-CarGame.SB2".

- Line 31 is triggered when the Car Sprite crosses the Finish Line which is denoted by "Red" pixels, the robot then plays Melody 2, displays "Success" for 1 sec. in the Stage Area, and then stops the Game. Hurrah for the user!

The Track Sprite has only one Script as shown in Fig. 2.64. It starts when it receives the "Start" message broadcasted by the Car Sprite earlier and it practically has the same functionality as the Main Script in "AD-Track.SB2" (Fig. 2.57) except for Line 4 (waiting for 5 seconds), which serves to synchronize the Track scrolling process to the countdown process being executed by the Car Sprite (Lines 10 through 17 in Fig. 2.61).

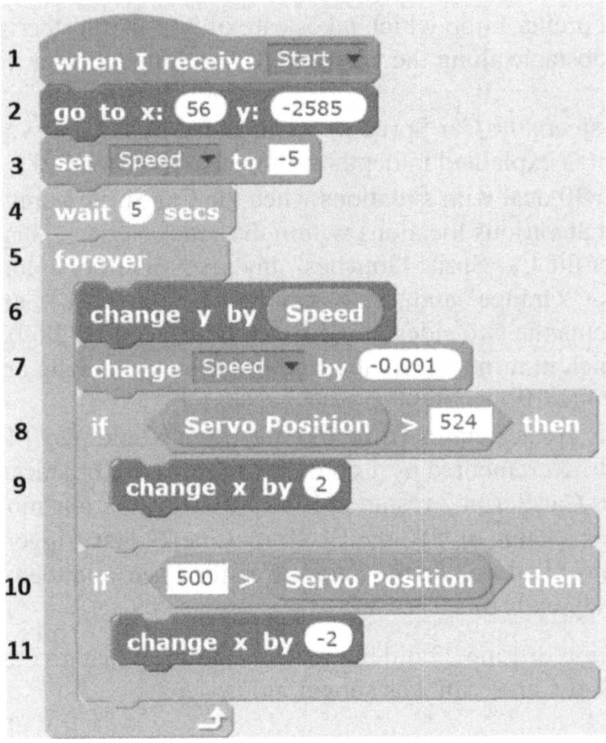

Fig. 2.64 Part 3 of Car Script 2 in Project "AD-CarGame.SB2".

Video 2.17 showcases the run-time performance for Project "AD-CarGame.SB2". The reader is encouraged to add "motor running" audio clips to this Scratch project, perhaps getting louder as the track rolling speed increases.

2.5 Using SCRATCH 3 with Edbot Dream Pro

Back in Spring 2019, SCRATCH 3 with Edbot 5 extensions was implemented as a web-enabled GUI, with the web server somewhere in the United Kingdom. This situation produces very slow sensor update capabilities for the DREAM robots. Thus, at the time of this book writing, the author is not recommending the use of SCRATCH 3 with DREAM robots if the SCRATCH project requires autonomous behaviors needing extensive access to "fresh" sensor data. Hopefully, the SCRATCH 3 DESKTOP version would become available in the near future – for updates please visit this web page www.cntrobotics.com/Scratch3.

For the reader's convenience, the author had converted some representative SCRATCH 2 example codes into their SCRATCH 3's equivalents:

- AD_Avoider-IF_ELSE_IF_1.SB3
- AD_Avoider-IF_ELSE_IF_2.SB3
- AD_Avoider-PAR_IF.SB3
- AD_Avoider-PAR_Loops.SB3
- AD_SensorThroughput.SB3
- AD_YinYang_TrackFollower_1.SB3
- AD_YinYang_TrackFollower_2.SB3
- AD_CarGame.SB3
- AD_ZeebZob_DowelSearch.SB3

The reader can draw his or her own conclusions regarding the use of SCRATCH 3 for their own needs.

2.6 Using PYTHON with Edbot Dream Pro

Using SCRATCH or PYTHON with the Edbot tool provides the same functionalities in controlling the typical Dream robot. However, with Python, the user can get one step closer to the "nuts and bolts" of the Edbot tool and learn some more details of the ROBOTIS firmware used on the Dream Controller CM-150. A PYTHON program will require more steps than a TASK or SCRATCH program to get the robot to do a typical action, but its execution speed is much higher and it will allow the user to tap into the vast resources of other PYTHON packages available at the Python Software Foundation (https://pypi.org/ and https://docs.python.org/3/contents.html). The web site Python Central is also a great resource for learning and practicing Python programming (https://www.pythoncentral.io/).

Edbot V.5 was a major upgrade which changed fundamentally its PYTHON API, thus all sample codes discussed in this section used the Edbot V.5.0.6.1280. The reader is assumed to have some basic skills in Python Programming, if not the author is recommending these books to get the reader better prepared for materials presented in this section (web resources abound for Python also):

- "Python for Tweens and Teens: Learn Computational And Algorithmic Thinking" (Bouras and Ainarozidou, 2017).
- "Python Crash Course", 2nd Edition (Matthes, 2019).

For more in-depth treatments of Python, the user is recommended to refer to these works:

- "Programming in Python 3", 2nd Edition (Summerfield, 2010).
- "The Python 3 Standard Library by Example", 1st Edition (Hellmann, 2017).

Documentation for the Edbot Dream Python API is available at this web link (http://support.ed.bot/edbot-dream-python3.html). This web page also shows how to install Python on the user's PC and provides a short primer on the programming steps needed for controlling the ROBOTIS Crocodile and Scorpion robots. When the Edbot tool was installed, it created a few Python example codes under the user's "Documents\Edbot\python" folder:

- dream_airplane.py
- dream_crocodile.py
- dream_elephant.py
- dream_motorcycle.py
- dream_scorpion.py

The author is using Thonny V.3.1.2 as the Integrated Development Environment (IDE) which is available for download at https://thonny.org/ , where it has a video demonstrating the basic usage of this IDE. Thonny V.3.1.2 also installs Python 3.7.2 for the user as part of its own installation and Thonny has its own GitHub page at https://github.com/thonny/thonny.

For the best use of the materials in Section 2.6, it is recommended that the reader opens the featured Python source code into Thonny and uses it while reading the rest of Section 2.6. In general, single-robot applications will use the robot "Zeeb", and multiple-robot applications will use the robots "Zeeb" and "Zob".

2.6.1 Sensors Throughput Applications

Let's get started with a basic Python program reading and displaying the current readings of the built-in Left-Center-Right IR sensors and the Ports 3 & 4 External IR Sensors IRSS-10 for the robot "Zeeb".

But before we can get into those gory details, we need to examine the data structure used by Edbot to keep track of the real-time values of **all its sensors** (built-in and external) **connected to each robot** configured in Edbot's setup.

Each robot's sensors-data structure is a Python dictionary named "reporters" (see link http://support.ed.bot/edbot-dream-python3.html#reporters), which is a list of paired "keys" and "numerical values" shown below as an example for robot "Zeeb":

```
"reporters": {
  "leftIR": 223,
  "centreIR": 14,
  "rightIR": 322,
  "clapCountLast": 0,
  "clapCountLive": 0,
  "port3": 580,
  "port4": 22,
  ...
}
```

Whereas, for example, the current raw value for the "leftIR" sensor is "223" and the current raw value for the "port3" external sensor (whatever it is) is valued at "580". "So far, so good", the reader must reckon at this point! But it is getting bad quickly as the next paragraph shows.

In order to get to this specific Zeeb's "reporters" dictionary from inside a Python program, we need to use a method named "get_data()" which will return a "huge" dictionary containing sensors data (+ other data structures) with the following keys:

robots	dict	The robots configured on the server. See the [robot] parameters dictionary above.
initComplete	boolean	Call to connect() has initialised.
server	dict	A dictionary containing the server **version** and **platform**.
auth	string	A unique token allocated by the server used to identify this session.
user	string	The current user connection.
users	array	The users connected to the server.

Unfortunately, the Zeeb's "reporters" dictionary, that we are interested in, is "buried" inside the "robots" dictionary which in turn is defined as follows:

model	dict	A dictionary containing the robot model **name**, **type** and **key**.
enabled	boolean	True if the robot is enabled.
connected	boolean	True if the robot is connected via Bluetooth.
activeUser	string	The currently active user. **None** means the active user is bypassed.
reporters	dict	A dictionary containing the Reporters such as sensor values.

This means that we need to use the value "Zeeb" and apply it to the "robots" dictionary in order to extract out the "reporters" dictionary of interest, which is then can be extracted further out for its individual "sensor" key and its corresponding numerical value! The implementation of this complex procedure will be shown later in Fig. 2.67.

For now, let's open the file "AD_SensorsThroughputBasic.py" into Thonny. Figs. 2.65 and 2.67 contain only the pertinent statements of this project to be explained further.

Fig. 2.65 displays the minimal steps to be used to ensure that connection and control have been properly established between Thonny, Edbot and the CM-150 controller. These statements are present in all example Python codes for this book:

```
7  import sys
8  import time
9  import edbot
14 ec = edbot.EdbotClient("localhost", 8080)
15 ec.connect()
20 robot1 = "Zeeb"
21 if not robot1 in ec.get_robot_names("dream"):
22     print(robot1 + " is not configured")
23     sys.exit()
28 print("Waiting to control " + robot1 + "... ", end="", flush=True)
29 ec.wait_until_active(robot1)
30 print("Got control of " + robot1 + "!")
```

Fig. 2.65 Steps to connect to robot and secure its control by Thonny via Edbot software.

- Line 7 imports the "sys" package so that we can use the methods exit() and print() at a later time.

- Line 8 imports the "time" package so that we can use the method time.sleep() later (Fig. 2.67).
- Line 9 imports the "edbot" package so that we can use its methods to control the Dream robot and to extract its sensors data.
- Line 14 uses the Edbot's method named "EdbotClient" to set up Thonny as a client to the Edbot software tool (which should be running already!) and assign this instance to Variable "ec". The argument "localhost" means that we are using the local PC as the Edbot server, and the argument "8080" specifies that Port 8080 is to be used. These settings need to match the settings used when the user installed and set up the Edbot package (see Fig. 2.66). Essentially, from the main menu bar of Edbot, if the user chooses Server >> Setup, then two options are available:
    - In "Settings", the user needs to make sure that the Server Port field is set to "8080".
    - In "Robots", the user needs to note the two options "Bypass active user" and "Available on network". The option, "Bypass active user", when checked, allows any connection to control the robot, i.e. both SCRATCH and THONNY can access the robot at the same time, so this option is needed when developing Python applications via Thonny. The option "Available on network" should be unchecked if the user is concerned about other users on the local network accessing inadvertently the robot that one is working on.

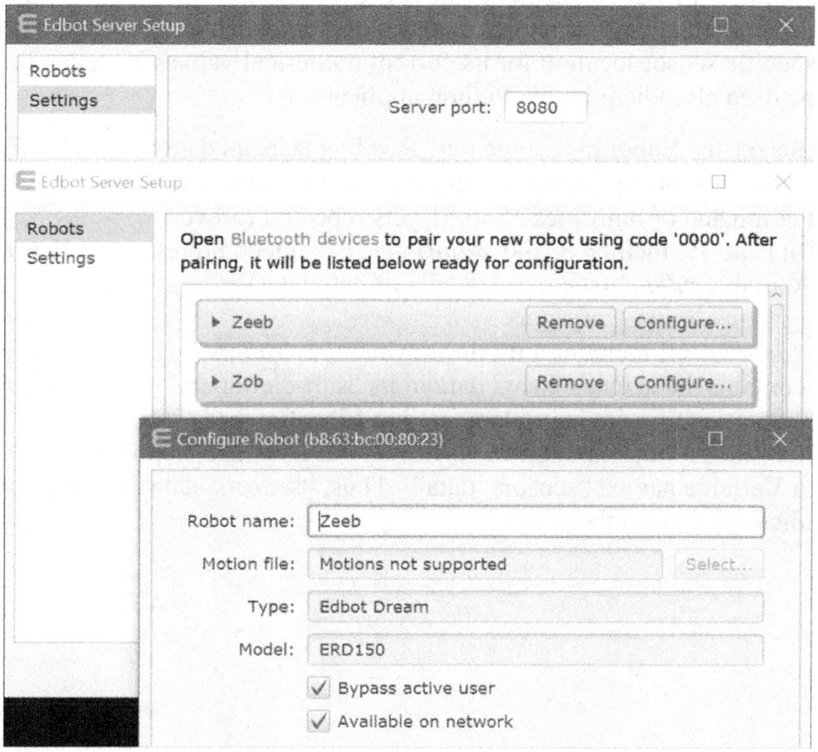

Fig. 2.66 Edbot Server Setup options for "Settings" and "Robots".

- Line 15 establishes the connection between Edbot and the user's Python session. Please note that the user can have **multiple instances** of Thonny accessing the same robot at any time.

- Line 20 assigns the string "Zeeb" to Variable "robot1". Of course, if the user's program needs to control a second robot, then the user needs to add a similar statement (***robot2 = "Zob"***) for example.
- Next, an IF structure (Lines 21-23) is used to check the **non-existence** of "robot1", i.e. string "Zeeb" in the setup configuration of the Edbot tool for "DREAM" type of robots, by calling the method ***get_robot_names("dream")***. If this condition happens to be True, the print statement in Line 22 gets executed, then this Python session is exited via method ***sys.exit()*** (Line 23). If this condition is False, meaning that "Zeeb" is properly configured in the Edbot tool, then program control flows to Line 28 which prints out ***"Waiting to control Zeeb…"***. The argument ***end=""*** means that the cursor should stay on the same print line (for the next print command in Line 30). Please note that by default the print() method goes to a new line after its execution, i.e. ***end="\n"***.
- Line 29 waits until "robot1" (i.e. Zeeb) gets active then program control flows to Line 30. Please note that if the option "Bypass active user" was not set properly earlier, Thonny would get stuck at Line 29 indefinitely.
- Line 30 prints out ***"Got control of Zeeb!"*** - on the same output line as for Line 29 (then it goes to a new line).

As previously explained, Fig. 2.67 illustrates the process of acquiring current (real-time) sensors data from the robot, via Edbot/Python data structures called dictionaries which can then be searched for a specific robot's name and a specific sensor location for its current numerical value which can be finally saved into a unique Variable, to be used elsewhere in this Python application:

- Line 33 makes the Edbot tool voice out "Reading sensors data from robot Zeeb".
- Line 34 is the Python equivalent for the start of the familiar Endless/Forever Loop. Thus, the block of commands from Lines 35 to 51 gets repeated forever:
    - In Line 35, method ***ec.get_data()*** gets executed and returns a **dictionary** saved into Variable ***"data"*** (see more details of get_data() at http://support.ed.bot/edbot-dream-python3.html#_get_data).
    - ***"data"*** itself includes a **dictionary** named ***"robots"*** which is of interest to us because it contains the actual sensors data in its "sub-dictionary" named ***"reporters"***. This double extraction process is implemented in Line 36 where "robot1" (i.e. "Zeeb") is used to go into the "robots" dictionary to extract out its entire "reporters" dictionary, and save it in a Variable named "sensors_data". Thus, "sensors_data" needs to be treated as a dictionary from then on.

```python
33 ec.say(robot1, "Reading sensors data from robot " + robot1)
34 while True:
35     data = ec.get_data()
36     sensors_data = data["robots"][robot1]["reporters"]
37     print("----------------------------------------")
38     if "leftIR" in sensors_data:
39         distance_raw_left = sensors_data["leftIR"]
40         print("Left IR = " + str(distance_raw_left))
41     if "centreIR" in sensors_data:
42         distance_raw_center = sensors_data["centreIR"]
43         print("Center IR = " + str(distance_raw_center))
44     if "rightIR" in sensors_data:
45         distance_raw_right = sensors_data["rightIR"]
46         print("Right IR = " + str(distance_raw_right))
47     if "port3" in sensors_data:
48         print("Port 3 IR = " + str(sensors_data["port3"]))
49     if "port4" in sensors_data:
50         print("Port 4 IR = " + str(sensors_data["port4"]))
51     time.sleep(0.05)
```

Fig. 2.67 Endless Loop used to extract Zeeb's selected sensors data from Edbot's data returned by method **get_data()**.

- Line 37 just prints out a separation line for each loop iteration.
- Lines 38-40 work together to check for the existence of the key "leftIR" in the dictionary "sensors_data" (Line 38), saves the corresponding numerical value into "distance_raw_left" (Line 39), and Line 40 prints out the message "Left IR = xxx" where xxx is the string-equivalent of the numerical variable "distance_raw_left".
- Lines 41-43 does the same tasks as described in the previous paragraph but for "distance_raw_center", and Lines 44-46 repeats similar tasks for "distance_raw_right".
- Lines 47-48 assumes that for whatever reason the user does not need to define an actual Variable for the key "port3", then the user can combine the extraction task inside the str() method to save 1 program statement. Lines 49-50 repeats the same procedure for the key "port4".
- Line 51 just pauses the loop for 0.05 second before looping back to Line 35.
- The user can try an experiment and comments out all the IF statements (Lines 36, 41, 44, 47, 49) to see any effect on the sensor data correctness and throughput. Anything of interest happened? (see "AD_SensorsThroughputBasic_NoIf.py").

Fig. 2.68 displays the Endless Loop of "AD_SensorsThroughputBasic_Zeeb.py" which uses the method get_robot() instead of get_data(). This version can be considered as a **streamlined** list of the required tasks. Please note that Line 35 in Fig. 2.68 shows that "ec.get_robot(robots1)" is equivalent to "data[robots][robot1]" in Line 36 of Fig. 2.67.

```
34 while True:
35     sensors_data = ec.get_robot(robot1)["reporters"]
36     print("-----------------------------------------")
37     print("Left IR = " + str(sensors_data["leftIR"]))
38     print("Center IR = " + str(sensors_data["centreIR"]))
39     print("Right IR = " + str(sensors_data["rightIR"]))
40     print("Port 3 IR = " + str(sensors_data["port3"]))
41     print("Port 4 IR = " + str(sensors_data["port4"]))
42     time.sleep(0.05)
```

Fig. 2.68 Endless Loop streamlined using method **get_robot(robot1),** i.e. **get_robot("Zeeb").**

Video 2.18 illustrates the run-time performances of these 3 versions of "AD_SensorsThroughputBasic". The increase in execution speed can be observed easily with the streamlining of the code.

Now, let's look at a 4th version named "AD_SensorsThroughput_Exceptions.py" to illustrate the handling of run-time exceptions - for more information please visit https://docs.python.org/3/tutorial/errors.html?highlight=exception. More details of the built-in exceptions can be found at https://docs.python.org/3/library/exceptions.html#bltin-exceptions.

We will be showcasing two types of exception: "KeyboardInterrupt", i.e. when the user pushes "Control-C" on the keyboard; and "KeyError", i.e. when a key, used in a dictionary, was not previously configured inside Edbot.

Fig. 2.69 shows the pertinent code segment whereas:

- Line 34 initializes Variable "counter" to 1 before entering the Endless Loop at Line 35.
- This EndLess Loop spans from Line 36 through Line 54, where a "try-except" structure is used to "capture" and "respond" to a specific run-time error:
    - The "try:" block contains all the statements needed to perform the needed tasks (Lines 37-48).
    - Lines 37-45 are for the tasks already explained in detail previously.
    - Line 46 increments Variable "counter" by 1, and if (counter == 101), updates Variable "robot1" with a new string "Zira" (Lines 47-48), else (i.e. counter <=100) loops back to Line 37. In other words, for the first 100 loop iterations, "robot1" is set to "Zeeb" which was well defined in Edbot's Setup – so no runtime errors. But at iteration 101, "robot1" is set to "Zira" (Line 48), which was not configured in Edbot's Setup previously, therefore this value change will cause a runtime error when the method ec.get_robot(robot1) gets called in the next loop iteration at Line 35, raising the exception "KeyError".
    - Next is the first "except:" block for the event "KeyboardInterrupt", i.e. when the user pushes "Control-C" on the keyboard. For this event, we want to print an appropriate error message before invoking "sys.exit()" to quit this program (Lines 49-51).
    - Last is the second "except:" block for the event "KeyError", i.e. when a dictionary is being accessed using a non-existent key value (for this example, this is when "Zira" is set – Lines 52-54).

```
# Read and display sensors data
ec.say(robot1, "Reading sensors data from robot " + robot1)
counter = 1
while True:
    try:
        sensors_data = ec.get_robot(robot1)["reporters"]
        print("----------------------------------------")
        print("Left IR = " + str(sensors_data["leftIR"]))
        print("Center IR = " + str(sensors_data["centreIR"]))
        print("Right IR = " + str(sensors_data["rightIR"]))
        print("Port 3 IR = " + str(sensors_data["port3"]))
        print("Port 4 IR = " + str(sensors_data["port4"]))
        print("Counter = " + str(counter))
        time.sleep(0.05)
        counter = counter + 1
        if (counter == 101):
            robot1 = "Zira"
    except KeyboardInterrupt:
        print("Session ended by user")
        sys.exit()
    except KeyError:
        print("Key = " + robot1 + "was not found")
        sys.exit()
```

Fig. 2.69 Pertinent code segment in "AD_SensorsThroughput_Exceptions.py".

Video 2.19 illustrates the run-time performance of "AD_SensorsThroughput_Exceptions.py" where the response to the exception KeyError was quite unexpected.

2.6.2 Avoider-Follower Applications

In this project "AD_Avoider_Follower_IfElseIf.py", we'll use the Pygame package to monitor the user's keyboard inputs and to provide graphical screen outputs at run time. The web site "https://www.pygame.org/docs/" has all the official PyGame documents that the reader can consult as needed. If the reader prefers a regular textbook, the author would recommend the work by Kinsley & McGugan (2015).

The Avoider robot named Zeeb is used in this project and see Fig. 2.70 for a screen capture at run time.

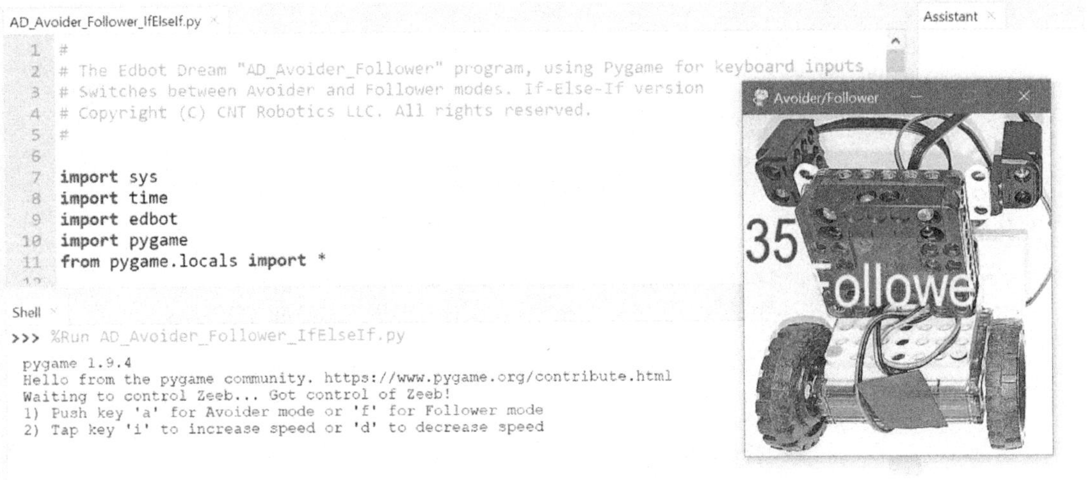

Fig. 2.70 Screen capture at run time for “AD_Avoider_Follower_IfElse_If.py”.

Figs. 2.71 through 2.75 display the pertinent code sections to be described in the following paragraphs.

```
7  import sys
8  import time
9  import edbot
10 import pygame
11 from pygame.locals import *
16 def forward():
17     global robot1, speed, r_speed
18     ec.set_servo_speed(robot1, "1/" + str(speed) + "/2/" + str(r_speed))
44 def inc_speed():
45     global speed, r_speed, robot1
46     if (speed < 95):
47         speed = speed + 5
48         r_speed = - speed
49         ec.set_buzzer(robot1, "1/3")
50
51 def dec_speed():
52     global speed, r_speed, robot1
53     if (speed > 5):
54         speed = speed - 5
55         r_speed = - speed
56         ec.set_buzzer(robot1, "3/3")
```

Fig. 2.71 Packages Import and Functions Definition for “AD_Avoider_Follower_IfElseIf.py”.

Lines 10-11 in Fig. 2.71 show how to import the Pygame package and a limited set of constants and functions that will be needed in this Python project (https://www.pygame.org/docs/tut/ImportInit.html).

In the source code for "AD_Avoider_Follower_IfElseIf.py", Lines 16-56 contain the user's definitions of functions used in this project: forward(), backward(), left(), left_back(), right(), right_back(), stop(), inc_speed() and dec_speed(), but Fig. 2.71 contains only the details for forward(), inc_speed() and dec_speed() which are described further below:

- Lines 16-18 shows the statements needed for Function forward() which sets the speeds for Motor "1" to "speed" and Motor "2" to "r_speed". Please note that as "speed" and "r_speed" are numerical variables, the str() functions are used to convert their numerical values to string equivalents before "joining" them into the final message to be sent to "robot1" (Line 18). Additionally, as Variables "robot1", "speed" and "r_speed" were first defined in the Main function, we need to let Function forward() know that these are "global" variables (Line 17).
- Lines 44-49 define Function inc_speed() which is used to increase the current value of Variable "speed" by 5 units (Line 47) if it is less than 95 (Line 46). This function finishes with updating Variable "r_speed" (Line 48) and setting off the buzzer with pitch "1" and for "0.3" second (Line 49) which is used as an audio feedback to let the user know that "speed" had been successfully increased. When the upper limit of 95 is attained, this function will not provide any further audio feedback.
- Lines 51-56 define Function dec_speed() which is used similarly to decrease the current value of Variable "speed" by 5 units (Line 54) if it is more than 5 (Line 53). This function finishes with updating Variable "r_speed" (Line 55) and setting off the buzzer with pitch "3" and for "0.3" second (Line 56) which is used as an audio feedback to let the user know that "speed" had been successfully decreased. When the lower limit of 5 is attained, this function will no longer provide any audio feedback.

Fig. 2.72 shows the steps needed to ensure a proper and working instance of an EdbotClient() and a set of constants to be used by Pygame:

- Lines 61-63 initialize Variables "robot1", "speed" and "r_speed" which are used by the EdbotClient instance.
- Lines 64-68 initialize constants needed for proper Pygame programming later:
    - Line 64 initializes "robot_mode" to "f" (i.e. Follower mode is default mode) and the user can change it later at runtime to "a" for Avoider mode.
    - Line 65 declares that the background image to be used is "Zeeb_m.jpg" and that its SCREEN_SIZE is 400x451 (Line 66) – a list with 2 elements.
    - Lines 67 and 68 define 2 messages to be used as a simple menu of options for the user at run time.
- Lines 75-76 create an instance of the EdbotClient class named "ec" and connect it to the Edbot Server software tool.
- Lines 80-82 check to see if "robot1" (i.e. "Zeeb") was properly configured in Edbot tool.
- Lines 86-88 wait for the active control of "robot1".
- Lines 89-90 print out the options menu for the user (see lower left corner of Fig. 2.70).

```
robot1 = "Zeeb"
speed = 35
r_speed = - speed
robot_mode = 'f' # a for Avoider, f for Follower
background_image_filename = 'Zeeb_m.jpg'
SCREEN_SIZE = (400, 451)
message1="1) Push key 'a' for Avoider mode or 'f' for Follower mode"
message2="2) Tap key 'i' to increase speed or 'd' to decrease speed"
ec = edbot.EdbotClient("localhost", 8080)
ec.connect()
if not robot1 in ec.get_robot_names("dream"):
    print(robot1 + " is not configured")
    sys.exit()
print("Waiting to control " + robot1 + "... ", end="")
ec.wait_until_active(robot1)
print("Got control of " + robot1 + "!")
print(message1)
print(message2)
```

Fig. 2.72 Setting up connection for EdbotClient and display Options Menu.

Fig. 2.73 shows how to activate Pygame and use some of its functionalities:

- "pygame" needs to be initialized as the first step (Line 94) (https://www.pygame.org/docs/tut/ImportInit.html).
- Line 95 pops up a window using SCREEN_SIZE to know how big to create it for and using 32-bit colors. Line 96 sets the caption "Avoider/Follower" as the title for this window (see top right area of Fig. 2.70). See (https://www.pygame.org/docs/ref/display.html?highlight=display%20set_mode#pygame.display.set_mode) and (https://www.pygame.org/docs/ref/display.html#pygame.display.set_caption).
- Line 97 loads the contents of the image file "Zeeb_m.jpg" into Variable "background" which is a Surface object (https://www.pygame.org/docs/ref/image.html?highlight=image%20load#pygame.image.load).
- Line 98 sets up Font Arial - Size 80 to be used for the two "text_surface" objects (little images) to be defined next.
- "text_surface1" is defined by Line 99 and contains the text "Avoider" to be rendered using a Yellow color by specifying its (Red, Green, Blue) components (i.e. 255 for Red, 255 for Green, and 0 for Blue). The argument "True" means that the rendered characters will have smooth edges (https://www.pygame.org/docs/ref/font.html?highlight=font%20render#pygame.font.Font.render).
- Lines 100-101 define the (x1, y1) coordinates where to put "text_surface1" on the previous "Avoider/Follower" window – essentially in the middle of the background image.
- Similarly, "text_surface2" (i.e. "Follower") would be rendered in Yellow and would be positioned in the middle of the background image also (Lines 102-104).

```python
pygame.init()
screen = pygame.display.set_mode(SCREEN_SIZE, 0, 32)
pygame.display.set_caption('Avoider/Follower')
background = pygame.image.load(background_image_filename).convert()
font = pygame.font.SysFont("arial", 80);
text_surface1 = font.render("Avoider", True, (255, 255, 0))
x1 = ( SCREEN_SIZE[0] - text_surface1.get_width() ) / 2
y1 = ( SCREEN_SIZE[1] - text_surface1.get_height() ) / 2
text_surface2 = font.render("Follower", True, (255, 255, 0))
x2 = ( SCREEN_SIZE[0] - text_surface2.get_width() ) / 2
y2 = ( SCREEN_SIZE[1] - text_surface2.get_height() ) / 2
```

Fig. 2.73 Setting up Pygame functionalities and "text_surfaces" 1 and 2.

```python
107 while True:
108 # Checking for user's keyboard inputs
109     for event in pygame.event.get():
110         if event.type == QUIT:
111             stop()
112             pygame.quit()
113             exit()
114 
115         if event.type == KEYDOWN:
116             if event.key == K_a:
117                 robot_mode = 'a'
118             if event.key == K_f:
119                 robot_mode = 'f'
120             if event.key == K_i:
121                 inc_speed();
122             if event.key == K_d:
123                 dec_speed();
124 
125     screen = pygame.display.set_mode(SCREEN_SIZE, 0, 32)
126     screen.blit(background, (0,0))
127     if (robot_mode == 'a'):
128         screen.blit(text_surface1, (x1, y1))
129     elif (robot_mode == 'f'):
130         screen.blit(text_surface2, (x2, y2))
131     text_surface3 = font.render(str(speed), True, (0, 0, 255))
132     screen.blit(text_surface3, (0, 120))
133     pygame.display.update()
```

Fig. 2.74 Monitoring keyboard events and displaying the "text_surface" objects.

Fig. 2.74 illustrates the first part of the Main Loop of this project where Pygame is used to monitor keyboard events and to display the popup window shown on the right side of Fig 2.70:

- Every time that the Main Loop is iterated (Line 107), Line 109 will get all the messages and remove them from the Pygame events queue via Function event.get() (https://www.pygame.org/docs/ref/event.html?highlight=event%20get#pygame.event.get).

- Line 109 also contains a For Loop that will check all the "events" received via Function event.get() for 2 specific event-types: "QUIT" and "KEYDOWN" (search web page https://www.pygame.org/docs/ref/event.html#pygame.event for a list of other event types).
    - If "QUIT" is found (Line 110), i.e. when the user closes the "Avoider/Follower" popup window, then the robot is "stopped" (Line 111), "pygame" is shut down (Line 112) and then the Python session is also exited (Line 113).
    - If "KEYDOWN" is found (Line 115), meaning that some keys had been pushed down, then the "event.key" variable is examined further (Lines 116-123):
        - If Key "a" had been pushed (Line 116), then "robot_mode" is set to "a" for Avoider mode (Line 117).
        - If Key "f" had been pushed (Line 118), then "robot_mode" is set to "f" for Follower mode (Line 119).
        - If Key "i" had been pushed (Line 120), then Function inc_speed() is called (Line 121).
        - Lastly, if Key "d" had been pushed (Line 122), then Function dec_speed() is called (Line 123).
- Lines 125-133 fill in the different display components to result in the popup window show in Fig. 2.70:
    - Line 125 initializes a window using SCREEN_SIZE, a "flag" value of 0 (i.e. a software-driven window), and a color depth of 32 bits and assigned this window object to Variable "screen" (https://www.pygame.org/docs/ref/display.html#pygame.display.set_mode).
    - Line 126 "blits" (i.e. draws) the "background" image defined back in Line 97 into the popup window (i.e. Surface object "screen").
    - Additionally, Lines 127-130 take care of "blitting" the appropriate "text_surface" object, depending whether "robot_mode" is in Avoider or Follower mode, right into the middle area of the object "screen".
    - Line 131 converts the numerical value of "speed" into a string which is then passed to Function "render" to make a "text_surface3" object with "blue" (0, 0, 255) characters. Line 132 next "blits" "text_surface3" onto "screen" at position (0, 120).
    - Finally, Function "display.update()" refreshes the popup window so that we can see the final result shown in Fig. 2.70 (https://www.pygame.org/docs/ref/display.html?highlight=display%20update#pygame.display.update).

Fig. 2.75 shows the statements to control the robot's motions depending on whether it is in "Avoider" or in "Follower" (default mode):

- Lines 137-139 extract the sensors data needed into "ir_left" and "ir_right".
- If the robot is in "Avoider" mode (Line 141), the control algorithm used is described in Lines 142 to 147. Essentially, if there is an obstacle on the left, the robot will perform a "right_back" maneuver, else if there is an obstacle on the right, it will perform a "left_back" maneuver, otherwise it will go forward. *Please note that there is another version of this project using a Parallel IFs control algorithm "AD_Avoider_Follower_ParIFs.py" and it is included in the sample codes for this book.*
- If the robot is in "Follower" mode (Line 149), the control algorithm used is described by Lines 150 to 157 whereas:

- If the object is real close to the IR sensors (Line 150), the robot will go forward to stay with the object motion (Line 151).
- Else if the object is detected on the left (Line 152), the robot will swing left (Line 153) to eventually face the object if it stays still.
- Similarly, the robot will swing to the right (Line 155) if it detects the object on the right (Line 154).
- If all other cases (Line 156), including when no object is detected, the robot stays still (Line 157).

```
136 #   Robot maneuvers start here
137     sensors_data = ec.get_robot(robot1)["reporters"]
138     ir_left = sensors_data["leftIR"]
139     ir_right = sensors_data["rightIR"]
140 # Avoider code
141     if (robot_mode == 'a'):
142         if (ir_left > 75):
143             right_back()
144         elif (ir_right > 75):
145             left_back()
146         else:
147             forward()
148 # Follower code
149     elif (robot_mode == 'f'):
150         if (ir_left > 250 or ir_right > 250):
151             forward()
152         elif (ir_left > 10):
153             left()
154         elif (ir_right > 10):
155             right()
156         else:
157             stop()
158 # End of Main
```

Fig. 2.75 Robot maneuvers control algorithms used in "AD_Avoider_Follower_IfElseIF.py".

Video 2.20 shows the runtime performance of these projects "AD_Avoider_Follower_IfElseIf.py" and "AD_Avoider_Follower_ParIFs.py". The author found that these projects would run fine first for a few minutes, and then Thonny would throw an internal error about WinError 10048: "Only one usage of each socket address (protocol/network address/port) is normally permitted". This error had been reported to Thonny developers so perhaps it will be resolved in the future. In the meantime, the author adds a small time delay for each Main Loop for 0.01 second, and so far it seems to take care of this issue (see sample Python files "AD_Avoider_Follower_IfElseIf_ts.py" and "AD_Avoider_Follower_ParIFs_ts.py").

2.6.3 Yin-Yang Track Follower

This project had been previously described in Section 2.4.6 for its SCRATCH implementation which was not very successful due to systemic delays in SCRATCH and in the sensor data flow. We shall see if this Python version would yield some performance improvements.

This Python project "AD_YinYang_TrackFollower.py" uses the same approach as the "Avoider_Follower" Python project so only the differing code sections will be described further in Figs. 2.76 to 2.80.

Fig. 2.76 displays the Variables used in this project and the changed or new ones are:

- "robot_mode" now can take on two values: "s" for stop mode and "t" for track mode (Line 50). So "message1" is changed accordingly (Line 56).
- "robot_move" is a new Variable which can be: "l" for left turn, "r" for right turn and "f" for forward (Line 51).
- The new background image is "YinYangTracker_s.jpg" and its SCREEN_SIZE is 451x436 (Lines 54 and 55).

```
49 robot1 = "Zeeb"
50 robot_mode = 's' # Initial mode is 'stop'
51 robot_move =''
52 speed = 35
53 r_speed = - speed
54 background_image_filename = 'YinYangTracker_s.jpg'
55 SCREEN_SIZE = (451, 436)
56 message1="1) Push key 't' to track or 's' to stop"
57 message2="2) Tap key 'i' to increase speed or 'd' to decrease speed"
```

Fig. 2.76 Global Variable definitions in "AD_YinYang_TrackFollower.py".

Fig. 2.77 lists the steps for initializing the PyGame package where 3 "text_surface" objects are now used "Forward" (Lines 88-90), "Left" (Line 91-93) and "Right" (Line 94-96), and like before, all 3 "text_surface" objects are to be centered on the background image when displayed.

```
83 pygame.init()
84 screen = pygame.display.set_mode(SCREEN_SIZE, 0, 32)
85 pygame.display.set_caption('YinYang Tracker')
86 background = pygame.image.load(background_image_filename).convert()
87 font = pygame.font.SysFont("arial", 80);
88 text_surface1 = font.render("Forward", True, (255, 255, 0))
89 x1 = ( SCREEN_SIZE[0] - text_surface1.get_width() ) / 2
90 y1 = ( SCREEN_SIZE[1] - text_surface1.get_height() ) / 2
91 text_surface2 = font.render("Left", True, (255, 255, 0))
92 x2 = ( SCREEN_SIZE[0] - text_surface2.get_width() ) / 2
93 y2 = ( SCREEN_SIZE[1] - text_surface2.get_height() ) / 2
94 text_surface3 = font.render("Right", True, (255, 255, 0))
95 x3 = ( SCREEN_SIZE[0] - text_surface3.get_width() ) / 2
96 y3 = ( SCREEN_SIZE[1] - text_surface3.get_height() ) / 2
```

Fig. 2.77 Setting up Caption and Text_Surfaces used in "AD_YinYang_TrackFollower.py".

Fig. 2.78 lists the PyGame display update scheme used:

- The new change is the IF_ELSE_IF structure (Lines 119-124) which is based on the current value of “robot_move” (Lines 119/121/123) and is used to display appropriately the 3 “text_surface” objects: “Forward” (Line 120), “Left” (Line 122) and “Right” (Line 124), and all 3 “text_surface” objects are centered on the background image when displayed.

```
117         screen = pygame.display.set_mode(SCREEN_SIZE, 0, 32)
118         screen.blit(background, (0,0))
119         if (robot_move == 'f'):
120             screen.blit(text_surface1, (x1, y1))
121         elif (robot_move == 'l'):
122             screen.blit(text_surface2, (x2, y2))
123         elif (robot_move == 'r'):
124             screen.blit(text_surface3, (x3, y3))
125         text_surface4 = font.render(str(speed), True, (0, 0, 255))
126         screen.blit(text_surface4, (0, 120))
127         pygame.display.update()
```

Fig. 2.78 Pygame actions for updating popup window for “AD_YinYang_TrackFollower.py”.

Fig. 2.79 lists the robot maneuvers control algorithm used:

- Four IR sensors are now used “ir_left”, “ir_right”, “port3_ir” and “port4_ir” (Lines 132-135).
- The same control algorithm explained in detail back in Fig. 2.37 is still used here, but it is slightly modified to account for the control values “s” or “t” for Variable “robot_mode” (Lines 138 and 154).
- If the robot is in “track” mode (robot_mode == ‘t’) then an IF-ELSE-IF structure (Lines 133-153) is used to determine the appropriate “Left” or “Right” moves for the robot depending where it is located: on “white” background (Lines 139-144) or on “black” background (Lines 145-150). For all remaining situations (Lines 151-153), the robot is commanded to go forward. The “robot_move” parameter is also adjusted accordingly so that the appropriate “text_surface” objects can be displayed (see Lines 119-123 of Fig. 2.78).
- If (robot_mode == ‘s’), Variable “robot_move” is cleared (Line 155) (so no text_surface will be displayed in the popup window) and the robot is also commanded to stop (Line 156).
- Finally, a time delay of 0.01 second is applied to each loop iteration to avoid the Windows Error 10048.

```
    sensors_data = ec.get_robot(robot1)["reporters"]
    ir_left = sensors_data["leftIR"]
    ir_right = sensors_data["rightIR"]
    port3_ir = sensors_data["port3"]
    port4_ir = sensors_data["port4"]

# Tracking the YinYang Line
    if (robot_mode == 't'):
        if (ir_left < 100 and port3_ir > 600 and port4_ir > 400):
            left()
            robot_move = 'l'
        elif (ir_right < 260 and port3_ir > 600 and port4_ir > 400):
            right()
            robot_move = 'r'
        elif (ir_left > 400 and port3_ir < 300 and port4_ir < 150):
            left()
            robot_move = 'l'
        elif (ir_right > 400 and port3_ir < 300 and port4_ir < 150):
            right()
            robot_move = 'r'
        else:
            forward()
            robot_move = 'f'
    elif (robot_mode == 's'):
        robot_move = ''
        stop()

    time.sleep(0.01)
# End of Main
```

Fig. 2.79 Robot maneuvers algorithm to follow an black-white altermating track in "AD_YinYang_TrackFollower.py".

Video 2.21 shows the runtime performance of the project "AD_YinYang_TrackFollower.py" which is much improved over the SCRATCH version "AD_YinYang_TrackFollower.SB2" in Section 2.4.6., although the TASK version "AD-YinYang-TrackFollower.TSK" is still the best performer because it runs locally on the CM-150.

2.6.4 Dowel Search Game using Threads & Events

This project had been previously described in Section 2.4.7 for its SCRATCH implementation which was using the concepts of Event Programming via "Message Broadcast" and "Message Received" between 3 independent Sprites (Zeeb, Zob and Cat as the Game Controller). The same Event Programming approach will be used in this Python project using Python-equivalent constructs which are "Thread" and "Event" objects from the "**threading**" module of the Python Standard Library (https://docs.python.org/3/library/threading.html#). The following web page was very helpful to the author by providing working code snippets used in this project: https://pymotw.com/3/threading/index.html#module-threading – it is maintained by Doug Hellmann.

This Python project "AD_ZeebZob_DowelSearch.py" uses the same approaches as previous Python projects regarding Edbot robot control and PyGame keyboard/graphics interfacing so only the newer concepts of threading and event signaling will be described further in the following paragraphs. If the reader is interested in textbooks to learn more about "threads" or "concurrency" in general, the author would suggest Hellmann (2017) and Nguyen (2018).

Fig. 2.80 shows all the import modules used in "AD_ZeebZob_DowelSearch.py".

```
import sys
import time
import random
import edbot
import pygame
from pygame.locals import *
import threading
```

Fig. 2.80 Modules imported in "AD_ZeebZob_DowelSearch.py".

In Python, a thread can be simply created as a regular function like the ones previously created for various robot maneuvers (https://en.wikibooks.org/wiki/Python_Programming/Threading). As we plan to use the Event Object formalism (https://docs.python.org/3/library/threading.html#event-objects), each thread will also need a special command called **message.wait()** which indefinitely blocks the execution of all codes following the **message.wait()** command. When that **message** is **set** in the main or in other threads, then the thread's code execution restarts and proceeds as normal for the rest of the thread/function. Thus the **message.wait()** command is the equivalent of the SCRATCH's HAT command "**When I receive Message**".

As an example, let's look at the details of Thread "zeeb_search()" (see Fig. 2.81):

- Line 49 lists the global variables used for this thread. "game_start" and "game_over" are Event objects, while "reset", "winner" and "robot1" are standard variables pre-defined in the Main Thread.
- Line 51 uses the command "game_start.wait()" which effectively blocks the execution of the remaining code for this thread which then can only wait for a signal "game_start.set()" from another thread (which will be the Main Thread for this case – see Fig. 2.85). Thus the signal **game_start.set()** is equivalent to the SCRATCH's **Broadcast Message** block.
- Once the signal "game_start.set()" is received by this thread, Lines 52-53 initializes the variables "dowel_distance_L" and "dowel_distance_R".
- Line 54 uses the random module to get a "1" or "2" to set the "start_direction" variable.
- Lines 55-56 are just trouble-shooting statements, the reader can comment them out if wanted.
- Lines 57-60 use an IF-ELSE structure to turn "robot1" (i.e. Zeeb) left or right depending on the randomized value for "start_direction".
- Lines 61-78 use an Endless Loop to implement the following search algorithm:
    - Lines 62-64 pick up the current values of the NIR sensors connected to Zeeb's Ports 3 and 4 using Edbot's Reporters dictionary for "robot1".
    - Lines 65-74 describes the details of an IF structure used to decide whether Zeeb can be declared a winner of a game session:
        - Line 65 indicates that once Zeeb finds an object close enough to either Port 3 or Port 4 sensor, it is commanded to go forward for 1 second.

```
47 # Thread 1 - zeeb_search
48 def zeeb_search():
49     global game_start, game_over, reset, winner, robot1
50 
51     game_start.wait()   # blocking until game_start is set in Main program
52     dowel_distance_L = 120
53     dowel_distance_R = 150
54     start_direction = random.randint(1,2)
55     print("\nZeeb's start direction = " + str(start_direction))
56     print("winner is " + str(winner))
57     if(start_direction == 1):
58             left(robot1)
59     else:
60             right(robot1)
61     while True:
62         sensors_data = ec.get_robot(robot1)["reporters"]
63         zeeb_port3_ir = sensors_data["port3"]
64         zeeb_port4_ir = sensors_data["port4"]
65         if (zeeb_port3_ir > dowel_distance_L or zeeb_port4_ir > dowel_distance_R):
66             forward(robot1)
67             time.sleep(1)
68             if (winner == -1):
69                 winner = 1 # Zeeb won!
70             else:
71                 winner = 0 # it's a draw
72             stop(robot1)
73             game_over.set()
74             break   # to break out of endless loop
75         if (reset == True):
76             stop(robot1)
77             break   # to break out of endless loop
78         time.sleep(0.01)  # to prevent Windows error 10048
79 # end of endless loop
80 
81     print("\nGetting out of zeeb_search()")
82     print("winner is " + str(winner))
83     return
```

Fig. 2.81 Thread/Function "zeeb_search()" in "AD_ZeebZob_DowelSearch.py".

- Lines 68-71 are used to decide whether Zeeb is the true winner (i.e. winner is set to "1", but if and only if winner was still set at "-1" from main thread), otherwise it is a draw and winner is set "0".
- Zeeb (robot1) is stopped (Line 72) and the signal game_over.set() (Line 73) is sent to all threads – in particular to the 3rd Thread game_end() (see Fig. 2.83).
- Then a break command is issued so that the thread transfer its execution to Lines 81-82 which are for informational use only and the Endless Loop is exited (Line 83).

- Lines 75-77 are used to take care of the situation when the user triggers a "reset" command from the main thread (see Fig. 2.85). Zeeb is then stopped (Line 76) and similarly the break command (Line 77) is used to exit the Endless Loop.
- For each iteration of the Endless Loop, Line 78 is used to slow down the process of using Edbot Reporters to read in sensors data (Line 62) so that a run-time error of the type 10048 won't be issued by the Windows OS.

Fig. 2.82 lists the details for Thread zob_search() which is virtually identical in design to Thread zeeb_search(), except that the global variable "robot2" is used and winner is set to "2" if Zob is the true winner of a game session.

```
85  # Thread 2 - zob_search
86  def zob_search():
87      global game_start, game_over, reset, winner, robot2
88
89      game_start.wait()   # blocking until game_start is set in Main program
90      dowel_distance_C = 110
91      start_direction = random.randint(1,2)
92      print("\nZob's start direction = " + str(start_direction))
93      print("winner is " + str(winner))
94      if(start_direction == 1):
95          left(robot2)
96      else:
97          right(robot2)
98
99      while True:
100         sensors_data = ec.get_robot(robot2)["reporters"]
101         zob_port3_ir = sensors_data["port3"]
102         zob_port4_ir = sensors_data["port4"]
103         if (zob_port3_ir > dowel_distance_C):
104             forward(robot2)
105             time.sleep(1)
106             if (winner == -1):
107                 winner = 2 # Zob won!
108             else:
109                 winner = 0 # it's a draw
110             stop(robot2)
111             game_over.set()
112             break  # to break out of forever loop
113         if (reset == True):
114             stop(robot2)
115             break  # to break out of foreever loop
116         time.sleep(0.01)   # to prevent Windows error 10048
117 # end of forever loop
118
119     print("\nGetting out of zob_search()")
120     print("winner is " + str(winner))
121     return
```

Fig. 2.82 Thread/Function "zob_search()" in "AD_ZeebZob_DowelSearch.py".

Fig. 2.83 lists the details for the third Thread game_end() which is designed to wrap up a game session once it receives the signal game_over.set() which can be sent by the Main Thread or by either Thread zeeb_search() (Line 73 of Fig. 2.81) or zob_search() (Line 111 of Fig. 2.82):

```
123 # Thread 3 - game_end()
124 def game_end():
125     global game_over, reset, robot1, robot2, winner
126 
127     game_over.wait()  # blocking until game_over is set in Main program or in the other 2 threads
128     stop(robot1)
129     stop(robot2)
130     ec.set_buzzer(robot1, "5/255")
131     ec.set_buzzer(robot2, "6/255")
132     if (winner == 1):
133         ec.say(robot1,"Zeeb won!")
134     elif (winner == 2):
135         ec.say(robot2,"Zob won!")
136     elif (winner == 0):
137         ec.say(robot2,"It is a draw")
138     elif (winner == -1):
139         ec.say(robot1,"Game time elapsed")
140     reset = True
141     print("\nGetting out of game_end()")
142     return
```

Fig. 2.83 Thread/Function "game_end()" in "AD_ZeebZob_DowelSearch.py".

- This thread uses the Event object "game_over.wait()" to block further execution of its remaining code (Line 127).
- Once the signal game_over.set() is sent from elsewhere, this thread immediately stops robot1 and robot2 (Lines 128-129) and makes them activate their buzzers (Lines 130-131).
- Lines 132-139 describes an IF-ELSE-IF structure designed to announce via Edbot's function "say" whether Zee or Zob has won, or that it is a draw, or if Variable "winner" is still at its initial value of "-1" then it means that the time allotted for one game session has elapsed.
- This thread's final actions are to set the Global Variable "reset" to TRUE and to print out an informational statement.

In short, creating a Python thread is no different than creating a regular Python function except for the execution control aspect which is provided by the signaling service of an Event object.

Next, let's look at how the various Thread and Event objects are defined and started, as shown in Fig. 2.84 (which are codes residing in the Main Thread):

- Lines 218 and 219 define the "threading.Event" objects used in this project which are "game_start" and "game_over". These Event objects manage an internal flag that is initially set to FALSE, which makes its corresponding method wait() block all code execution flow wherever they are used. Code execution would resume as normal when this flag is set to TRUE somewhere else by using its corresponding method set(). This web page https://docs.python.org/3/library/threading.html#event-objects documents other parameters that can be applied depending on the user's needs.
- Line 222 associates Thread "t1" with the Target Function "zeeb_search" and Line 223 starts "t1", i.e. Function zeeb_search() is called, but only Line 49 of Fig. 2.81 really gets executed.

Next, Line 51, with its game_start.wait() method, effectively holds this thread in hiatus until game_start.set() is used somewhere else at a later time.

```
217 # Define the two thread controling events used
218 game_start = threading.Event()
219 game_over = threading.Event()
220
221 # Define the three threads used
222 t1 = threading.Thread(name='Zeeb Search', target=zeeb_search, args=() )
223 t1.start()
224
225 t2 = threading.Thread(name='Zob Search', target=zob_search, args=() )
226 t2.start()
227
228 t3 = threading.Thread(name='End Game', target=game_end, args=() )
229 t3.start()
```

Fig. 2.84 Thread and Event objects definitions and starting actions in "AD_ZeebZob_DowelSearch.py".

- Lines 225-226 do a similar job of association of "t2" to Target Function "zob_search" and for starting "t2". And in the same way as for "t1", "t2" halts its execution flow at Line 89 of Fig. 2.82 and waits for the signal game_start.set() to be activated somewhere else at a later time.
- Lastly, Lines 228-229 start the third thread "t3", i.e. Function game_end(), which also halts its execution flow at Line 127 of Fig. 2.83 and waits for the signal game_end.set() to be activated elsewhere.

If the reader goes back and pulls up the SCRATCH code "AD_ZeebZob_DowelSearch.sb2", the reader can see that so far for Python we have kept the same code design approach as in SCRATCH: with Sprite "Zeeb" corresponding to Thread "zeeb_search()" and Sprite "Zob" corresponding to Thread "zob_search()". The Thread "game_end()" corresponds to the SCRATCH "Game Over" block which is located with the "Cat" Sprite, but it could have been put in either the Zeeb or Zob Sprites and the run-time functionality would be the same. Let's next see how we can implement the "Cat" Sprite to act as the Game Controller in Python.

Fig. 2.85 shows the start of the Endless Loop that acts as the Game Controller (which resides in the Main Thread):

- Line 232 initializes Variable "time_game_start" to 0, but it won't actually be set until the user pushes on the key "p" during run-time.
- The Endless Loop starts at Line 233, and the first task is to process the PyGame Event Queue with a For Loop defined by Lines 235-263, looking for 2 event types – QUIT or KEYDOWN:
    - If the event type is QUIT (Lines 236-240), i.e. the user chooses to close the "Zeeb-Zob Game" window as shown in Fig. 2.86, then both robots are stopped (Lines 237-238), the PyGame interface closes (Line 239) and the Python session is also exited (Line 240).
    - If the user pushes down on key "p", then the "Game" starts with the sending of a TRUE signal by "game_start.set()" (Line 244) to release the blocks set by the methods "game_start.wait()" used in the Threads "zeeb_search()" (Line 51 of Fig. 2.81) and "zob_search()" (Line 89 of Fig. 2.82). At this time the user would see the robots spinning and looking for an object as the full codes for "zeeb_search()" and

"zob_search()" are finally doing their jobs! Line 245 marks the effective "time_game_start" using the method process_time() from the time module. This parameter is also printed out on Thonny's Shell for informational purposes (Line 246). Lines 247-248 "screen.blit" "text_surface1" which is "Start" in the middle of the window named "Zeeb-Zob Game" (see Fig. 2.86). Next "time.sleep()" of 1 second is used to display the "Start" message hopefully long enough for the user to notice it.

```
231 # Main controller loop starts here
232 time_game_start = 0.0  # initial value, actual value is set by event.key "p"
233 while True:
234 # Checking for user's keyboard inputs
235     for event in pygame.event.get():
236         if event.type == QUIT:
237             stop(robot1)
238             stop(robot2)
239             pygame.quit()
240             sys.exit()  # sys.exit() when user closes Pygame window
241 
242         if event.type == KEYDOWN:
243             if event.key == K_p:
244                 game_start.set()  # signals zeeb_search and zob_search to run
245                 time_game_start = time.process_time()   # the effective time_game_start
246                 print("time_game_start = " + str(time_game_start))
247                 screen.blit(text_surface1, (x1, y1))
248                 pygame.display.update()
249                 time.sleep(1)
250             if event.key == K_s:
251                 stop(robot1)
252                 stop(robot2)
253                 reset = True  # this flag is also used to control threads behavior
254                 screen.blit(text_surface5, (x5, y5))
255                 pygame.display.update()
256                 time.sleep(2.0)
257                 sys.exit()  # sys.exit() when user wants to stop the game
258             if event.key == K_i:
259                 inc_speed(robot1)
260                 inc_speed(robot2)
261             if event.key == K_d:
262                 dec_speed(robot1)
263                 dec_speed(robot2)
```

Fig. 2.85 Part 1 of Main Controller Loop in "AD_ZeebZob_DowelSearch.py".

o If the user pushes down on key "s", then the "Game" exits by first stopping both robots (Lines 251-252). Next, the global parameter "reset" is set to TRUE to signal the Threads "zeeb_search()" and "zob_search()" to stop the respective robots (once more – just to be sure!) and also to force a break from the Endless Loops that both threads were in at this point in time (see Lines 75-77 of Fig. 2.81 and Lines 113-115 of Fig. 2.82) and thus these threads/functions are terminated properly and cleanly. This is an important step to remember to do when ones use multiple threads in an application. Lines 254-256 display "text_surface5", i.e. "Game Stopped" for 2 seconds, before a system exit is executed at Line 257.

o If the user pushes down on key “i” the robots speeds are increased by 5 units (Lines 258-260) using Function inc_speed() (see Fig. 2.87).

o And if the user pushes down on key “d” the robots speeds are decreased by 5 units (Lines 261-263) using Function dec_speed() (also see Fig. 2.87).

AD_ZeebZob_DowelSearch.py

```
#
# The Edbot Dream "AD_ZeebZob_DowelSearch" program, using Pygame for keyboard inputs
# Using threads and event objects for signaling scheme
# Copyright (C) 2019 - CNT Robotics LLC. All rights reserved.
#

import sys
import time
import random
import edbot
import pygame
from pygame.locals import *
import threading

#
# Functions definitions start here
#
def forward(robot_name):
    global speed, r_speed
    ec.set_servo_speed(robot_name, "1/" + str(speed) + "/2/" + str(r_speed))

```

Shell

```
Getting out of zob_search()
winner is 2

Getting out of game_end()

Getting out of zeeb_search()
winner is 2
```

Fig. 2.86 Runtime Display for “AD_ZeebZob_DowelSearch.py”.

```
def inc_speed(robot_name):
    global speed, r_speed
    if (speed < 95):
        speed = speed + 5
        r_speed = - speed
        ec.set_buzzer(robot_name, "1/3")

def dec_speed(robot_name):
    global speed, r_speed
    if (speed > 5):
        speed = speed - 5
        r_speed = - speed
        ec.set_buzzer(robot_name, "3/3")
```

Fig. 2.87 Functions inc_speed() and dec_speed() in “AD_ZeebZob_DowelSearch.py”.

Fig. 2.88 shows Part 2 of the Endless Loop that acts as the Game Controller (which resides in the Main Thread). It is designed to update the user about the progress of the game via the "Zeeb-Zob Game" window as shown in Fig. 2.86:

```
265        screen = pygame.display.set_mode(SCREEN_SIZE, 0, 32)
266        screen.blit(background, (0,0))
267        if (winner == 1):
268            screen.blit(text_surface2, (x2, y2))
269            time.sleep(5.0)
270        elif (winner == 2):
271            screen.blit(text_surface3, (x3, y3))
272            time.sleep(5.0)
273        elif (winner == 0):
274            screen.blit(text_surface4, (x4, y4))
275            time.sleep(5.0)
276        text_surface7 = font.render(str(speed), True, (0, 0, 255))
277        screen.blit(text_surface7, (240, 0))
278        pygame.display.update()
```

Fig. 2.88 Part 2 of Main Controller Loop in "AD_ZeebZob_DowelSearch.py".

- Line 265 opens a window and Line 266 blits the background image into it (i.e. the "Zeeb-Zob Game" window).
- The IF-ELSE-IF structure (Lines 267-275) takes care of displaying the "winner": Zeeb (winner == 1) or Zob (winner == 2) for each game session, or declares that it is a draw (winner == 0).
- Line 276 assigns the current value of Variable "speed" to "text_surface7" in blue characters and Line 277 blits it into the "Zeeb-Zob Game" window opened by Line 265.
- Line 278 updates the "Zeeb-Zob Game" window with all the appropriate changes, background image combined with various text messages, so that the user can see them.

Fig. 2.89 lists Part 3 of the Main Controller Loop which, for each loop iteration, checks to see whether the game "elapsed" time so far has gone beyond **3 seconds** based on the "elapsed" process_time() for the Main Thread since "time_game_start" was set back on Line 245 of Fig. 2.85:

- If this condition is True, then the signal game_over.set() (Line 282) is sent to Thread 3, i.e. Function game_end(), to make it resume its execution at Line 128 of Fig. 2.83 until the return statement at Line 142 which essentially makes Edbot voice out the winning robot or that it was a draw (this is done in addition to the display of the appropriate texts as programmed in Part 2 of the Main Controller Loop in Fig. 2.88).
- Line 283 is important as "reset" is used to terminate Threads "zeeb_search()" and zob_search().
- Lines 284-286 refresh the "Zeeb-Zob Game" window with the background image and "text_surface6" which is the message "Over Time Limit".
- Line 287 breaks the Main Controller Loop, and after 5 seconds (Line 290), the Python session itself ends with sys.exit() at Line 291.

```
    if ( (time_game_start > 0.0) and (time.process_time() - time_game_start) >= 3.0 ):
        print("time_game_elapsed = " + str((time.process_time() - time_game_start)))
        game_over.set()
        reset = True
        screen.blit(background, (0,0))
        screen.blit(text_surface6, (x6, y6))
        pygame.display.update()
        break   # to break out of main endless loop
# end of Main endless loop

time.sleep(5)  # so that user can see the display of final messages 5 seconds longer
sys.exit()   # Main sys.exit()
# End of Main
```

Fig. 2.89 Part 3 of Main Controller Loop in "AD_ZeebZob_DowelSearch.py".

Video 2.22 shows the runtime performance of the project "AD_ZeebZob_DowelSearch.py" which interestingly shows that the process_time() method does not actually measure the absolute physical time as this Python program cannot control how the Windows OS allocate its computing resources at run time.

In this Section 2.6.4, the author hopes to have illustrated how to use Python threads and events to add "process concurrency" to the multi-robots control methods already provided by the Edbot tool.

2.6.5 Using USB Camera with OpenCV & NumPy

Another advantage when using Python which is a widely adopted language is the large repository of modules/tools found at the Python Software Foundation https://pypi.org/. This section 2.6.5 would illustrate the use of the OpenCV module and a related module called NumPy (https://pypi.org/project/opencv-python/). If the reader is using Thonny, it is recommended to use the "Manage Package" tool to search for the module "opencv-python" and install it. This process will install all related modules for the reader in one step. The author is using OpenCV-Python V.4.1.0.25 and a Logitech webcam C910 for the example programs in this section (see Fig. 2.90).

Fig. 2.90 Zeeb robot with webcam C910.

The field of Computer Vision or Machine Vision is large and complex, thus in this section the author is only trying to show that, with a bit of work, the reader can incorporate Machine Vision capabilities to a "lowly" DREAM robot. References for Machine Vision, OpenCV and NumPy abound, so the author is only recommending a few that he is familiar with:

- "Learning OpenCV 3" by Kaehler and Bradski (2017), this book is written for the C/C++ programming language but it is a thorough reference to keep.
- For Cookbook types of OpenCV book references in Python, the author has used "Learn OpenCV 4 by Building Projects" by Escrivá and Mendonça (2018) and "OpenCV 3 Computer Vision with Python Cookbook" by Spizhevoy and Rybnikov (2018) and had found them useful.
- Web resources (free and at cost) for OpenCV-Python and NumPy can be found at many places and here are a few:
    - https://opencv-python-tutroals.readthedocs.io/en/latest/index.html
    - https://www.learnopencv.com/about/
    - https://www.pyimagesearch.com/
    - http://www.numpy.org/ and https://docs.scipy.org/doc/numpy/.
    - http://cs231n.github.io/python-numpy-tutorial/

The author assumes that most readers are aware that a color digital image consists of physical pixels in a 2-dimensional array such as 1280 x 1024 or 3840 x 1600 (https://en.wikipedia.org/wiki/Color_image). Each of these physical pixels is in fact a triplet of pixels in the primary colors Red, Green, and Blue (also known as the RGB Color Model - https://en.wikipedia.org/wiki/RGB_color_model). The RGB Color Model is needed to capture a color scene from the real world into a video camera sensor and to display the same color scene on a computer monitor (https://www.visiononline.org/userassets/aiauploads/file/cvp_the-fundamentals-of-camera-and-image-sensor-technology_jon-chouinard.pdf). When these triplet light sources from a color display reach the human eye, a very complex bio-physical process occurs to give us the sensation of color (https://www.olympus-lifescience.com/en/microscope-resource/primer/lightandcolor/humanvisionintro/). In turn, we usually describe those color sensations as red, green and blue (the primary colors), but also as yellow or brown or violet (generally called "hue"). For example, to create the "yellow" sensation we would have to turn on the "red" and "green" light sources while providing no "blue" light. Thus, from the point of view of color image processing, we would have to use at least two parameters (Red & Green) to define a single hue (Yellow). For this reason, other color models are created (derived from the RGB model) to more closely align with how humans classify perceived colors (https://en.wikipedia.org/wiki/HSL_and_HSV), where we would need only one parameter to describe Hue (with the other two parameters S or L/V used to describe more subtle color perceptions such as vividness or brightness of a particular color perceived). We will be using the HSV Color Model in the Python projects described in this Section 2.6.5.

First let's look at the project "AD_RemoteVision_HSV.py" where the user can remotely control the movement of Zeeb (therefore the camera – see Fig. 2.90) to position a central Region Of Interest (ROI) onto an object of interest to display its HSV data (see Fig. 2.91). This project also gives the user the option of recording and pausing multiple video clips into a single video file.

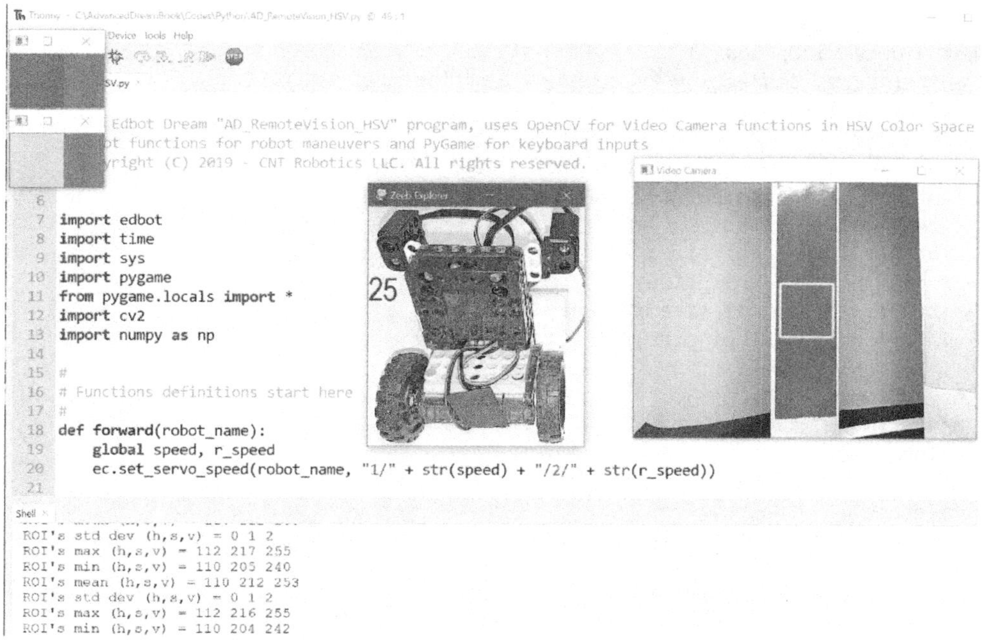

Fig. 2.91 Run-time Displays for "AD_RemoteVision_HSV.py".

Fig. 2.92 shows the modules imported for this project, please note "cv2" and "numpy".

```
import edbot
import time
import sys
import pygame
from pygame.locals import *
import cv2
import numpy as np
```

Fig. 2.92 Imported modules in "AD_RemoteVision_HSV.py".

Fig. 2.93 shows the main Variables used and the Messages for the run-time options menu:

```
robot1 = "Zeeb"
robot_mode = 's' # Initial mode is 'stop'
speed = 35
r_speed = - speed
message1 = "1) Use Arrow keys to move robot around or 's' to stop the robot"
message2 = "2) Tap key 'i' to increase speed or 'd' to decrease speed"
message3 = "3) Tap key 'r' to toggle start/stop video and 'Esc' to quit this app"
```

Fig. 2.93 Initialization of Variables and Constants used in "AD_RemoteVision_HSV.py".

Fig. 2.94 illustrates the various initialization and checking steps that need to be done in order to use a video camera as a VideoCapture object and a VideoWriter object to record video clips selected from the video stream:

```
89  vid_cam = cv2.VideoCapture(0)
90  if(not vid_cam.isOpened()):
91      print("Error connecting to camera - Exiting program")
92      vid_cam.release()
93      sys.exit()
94  f_width = int(vid_cam.get(cv2.CAP_PROP_FRAME_WIDTH))
95  f_height = int(vid_cam.get(cv2.CAP_PROP_FRAME_HEIGHT))
96  roi_size = 100
97  w1 = int((f_width - roi_size)/2)
98  h1 = int((f_height - roi_size)/2)
99  w2 = int((f_width + roi_size)/2)
100 h2 = int((f_height + roi_size)/2)
101 # in this version we use HSV Color Space for the ROI
102 min_hsv = [None] * 3
103 max_hsv = [None] * 3
104 recording = False
105
106 vid_filename = "video.avi"
107 vid_recorder = cv2.VideoWriter(vid_filename, cv2.VideoWriter_fourcc('M','J','P','G'),
108                                25, (f_width, f_height))
109 if(not vid_recorder.isOpened()):
110     print("Error opening video recording file - Exiting program")
111     vid_recorder.release()
112     sys.exit()
```

Fig. 2.94 Setting up VideoCapture and VideoWriter objects in "AD_RemoteVision_HSV.py".

- Line 89 associates VideoCapture's Device 0 with the object named "vid_cam" (https://docs.opencv.org/4.1.0/d8/dfe/classcv_1_1VideoCapture.html).
- Line 90-93 checks for proper connection to the camera by checking on method "isOpened()", and if there is a connection error, the program releases the "vid_cam" resource and exits promptly at Line 93. If not, the program proceeds on to Line 94.
- Lines 94-104 lists the initializations of various Variables and Constants used in the program:
  - Please note that the ROI is a square of 100x100 pixels and (w1, h1) and (w2, h2) denote respectively its top left and bottom right corners in relation to the video frame coordinates system (f_width, f_height). In fact, the ROI is centered with respect to the video frame (see yellow square in Fig. 2.91).
  - "min_hsv" and "max_hsv" are 1-dimensional arrays with 3 elements having zero for initial values.
- Line 106 assigns a local file name "video.avi" to Variable "vid_filename".
- In Line 107, the class "VideoWriter()" creates a "vid_recorder" object using 4 arguments:
  - "vid_filename", i.e. the string "video.avi".
  - a FOURCC value of "MJPG", specifying the video codec to use when saving the video frames (see http://www.fourcc.org/codecs.php for other codecs that the reader can try out). At this point, some readers may ask why did the author use the file extension of "avi" for "vid_filename" in Line 106, instead of "mp4"? The reason was that so that most Windows video utilities like "Windows Media Player" or "Movies & TV" can play this video file. When the author changed the file extension to "mp4", then only "VLC Media Player" or "5KPlayer" can play such file, and the author so far could not find any technical reason for this effect!
  - "25" means 25 video frames per second.

- The pair (f_width, f_height) means that the original video frame size streaming in from the camera 0 is kept (see Lines 94 and 95).

- Lines 109-112 checks to see if the "vid_recorder" object can be opened properly, if not the whole program exits at Line 112. If "vid_recorder" opens properly, then the program flows into the code segment displayed in Fig. 2.95.

Fig. 2.95 defines the main PyGame window named "Zeeb Explorer" with its background image "Zeeb_m.jpg" and various Text Surfaces to be used in later parts of the program.

```
pygame.init()
background_image_filename = 'Zeeb_m.jpg'
SCREEN_SIZE = (400, 451)
screen = pygame.display.set_mode(SCREEN_SIZE, 0, 32)
pygame.display.set_caption('Zeeb Explorer')
background = pygame.image.load(background_image_filename).convert()
font = pygame.font.SysFont("arial", 60);
text_surface1 = font.render("Forward", True, (255, 255, 0))
x1 = ( SCREEN_SIZE[0] - text_surface1.get_width() ) / 2
y1 = ( SCREEN_SIZE[1] - text_surface1.get_height() ) / 2
text_surface2 = font.render("Left", True, (255, 255, 0))
x2 = ( SCREEN_SIZE[0] - text_surface2.get_width() ) / 2
y2 = ( SCREEN_SIZE[1] - text_surface2.get_height() ) / 2
text_surface3 = font.render("Right", True, (255, 255, 0))
x3 = ( SCREEN_SIZE[0] - text_surface3.get_width() ) / 2
y3 = ( SCREEN_SIZE[1] - text_surface3.get_height() ) / 2
text_surface4 = font.render("Backward", True, (255, 255, 0))
x4 = ( SCREEN_SIZE[0] - text_surface4.get_width() ) / 2
y4 = ( SCREEN_SIZE[1] - text_surface4.get_height() ) / 2
text_surface5 = font.render("Stopped", True, (255, 0, 0))
x5 = ( SCREEN_SIZE[0] - text_surface5.get_width() ) / 2
y5 = ( SCREEN_SIZE[1] - text_surface5.get_height() ) / 2
text_surface6 = font.render("Recording", True, (255, 0, 255))
x6 = ( SCREEN_SIZE[0] - text_surface6.get_width() ) / 2
y6 = 350
text_surface7 = font.render("Pause Record", True, (255, 0, 255))
x7 = ( SCREEN_SIZE[0] - text_surface7.get_width() ) / 2
y7 = 350
```

Fig. 2.95 Setting up various PyGame objects in "AD_RemoteVision_HSV.py".

Fig. 2.96 displays Part 1 of the Main Endless Loop which checks for the PyGame event type "QUIT", i.e. when the user closes the window named "Zeeb Explorer" in Fig. 2.91:

- Line 152 stops the robot.
- Lines 153-155 close all OpenCV devices and windows used in the program.
- Line 156 quits the PyGame module and Line 157 exits the Python shell.

```
148 while True:
149 # Checking for user's keyboard inputs
150     for event in pygame.event.get():
151         if event.type == QUIT:  # when user closes Pygame window
152             stop(robot1)
153             vid_cam.release()
154             vid_recorder.release()
155             cv2.destroyAllWindows()
156             pygame.quit()
157             sys.exit()
```

Fig. 2.96 Part 1 of Main Endless Loop in "AD_RemoteVision_HSV.py".

Fig. 2.97 displays Part 2 of the Main Endless Loop which checks for the PyGame event type "KEYDOWN", i.e. when the user pushes on specific keys:

```
158         if event.type == KEYDOWN:
159             if (event.key == K_r):  # toggling start/stop video recording
160                 if (recording == False):
161                     recording = True
162                     screen.blit(text_surface6, (x6, y6))
163                     pygame.display.update()
164                 else:
165                     recording = False
166                     screen.blit(text_surface7, (x7, y7))
167                     pygame.display.update()
168             if event.key == K_s:  # stop the robot
169                 robot_mode = 's'
170                 stop(robot1)
171             if event.key == K_i:  # increase robot speed
172                 inc_speed(robot1)
173             if event.key == K_d:  # decrease robot speed
174                 dec_speed(robot1)
175             if event.key == K_UP:  # robot forward
176                 robot_mode = 'f'
177                 forward(robot1)
178             if event.key == K_DOWN:  # robot backward
179                 robot_mode = 'b'
180                 backward(robot1)
181             if event.key == K_LEFT:  # robot left
182                 robot_mode = 'l'
183                 left(robot1)
184             if event.key == K_RIGHT:  # robot right
185                 robot_mode = 'r'
186                 right(robot1)
```

Fig. 2.97 Part 2 of Main Endless Loop in "AD_RemoteVision_HSV.py".

- Lines 159-167 illustrates when the user toggles on key "r". Essentially, if "recording" was not in effect (Line 160) then "recording" is set to True (Line 161) and "text_surface6", i.e.

"Recording", is displayed (Lines 162-163). Otherwise set "recording" to False (Line 165) and display "text_surface7", i.e. "Pause Record" (Lines 166-167).

- Lines 168-186 set the proper value to "robot_mode" and effectively perform various robot maneuvers as commanded by the user: "stop", "increase speed", "decrease speed", "forward", "backward", "left" and "right" (using keys "s", "i" and "d", and the respective arrow keys).

Fig 2.98 displays Part 3 of the main loop where PyGame blits the proper text surfaces depending on the current values of parameters "robot_mode", "recording", and the current speed setting (Lines 204-205).

```
188        screen = pygame.display.set_mode(SCREEN_SIZE, 0, 32)
189        screen.blit(background, (0,0))
190        if (robot_mode == 'f'):    # updating robot motion directions
191            screen.blit(text_surface1, (x1, y1))
192        elif (robot_mode == 'l'):
193            screen.blit(text_surface2, (x2, y2))
194        elif (robot_mode == 'r'):
195            screen.blit(text_surface3, (x3, y3))
196        elif (robot_mode == 'b'):
197            screen.blit(text_surface4, (x4, y4))
198        elif (robot_mode == 's'):
199            screen.blit(text_surface5, (x5, y5))
200        if (recording == True):    # updating video recording operations
201            screen.blit(text_surface6, (x6, y6))
202        elif (recording == False):
203            screen.blit(text_surface7, (x7, y7))
204        text_surface8 = font.render(str(speed), True, (0, 0, 255))    # updating current robot speed
205        screen.blit(text_surface8, (0, 120))
206        pygame.display.update()
```

Fig. 2.98 Part 3 of Main Endless Loop in "AD_RemoteVision_HSV.py".

Fig. 2.99 shows Part 4 where for each iteration of the Main Endless Loop a video frame is captured (Line 208) and checked for proper capture (Line 209), if an error occurred then break out of the Main Loop (Lines 210-211). Otherwise check to see if "recording" is wanted (Line 212) and if "recording" is True save the current video frame using method "vid_recorder.write()" (Line 213).

```
207 # Operating video camera
208     got_frame, vid_frame = vid_cam.read()  # picking up a new video frame
209     if not got_frame:    # error capturing video frame
210         print("No video frame captured!")
211         break  # get out of Main Loop
212     if (recording == True):  # saving video frame per user's toggling the 'r' key
213         vid_recorder.write(vid_frame)
```

Fig. 2.99 Part 4 of Main Endless Loop in "AD_RemoteVision_HSV.py".

A special note should be mentioned for Line 208 which shows the compactness of the Python language where two items get returned whenever method "vid_cam.read()" is invoked: "got_frame" is a logical flag while "vid_frame" is a 3-dimensional array representing the image captured in RGB Color Model (or to be precise in BGR format – a peculiar quirk of OpenCV). "vid_frame" will be further processed in Part 5.

Fig. 2.100 shows Part 5 where the true powers of OpenCV and NumPy are displayed:

- Line 215 is quite a "nifty" NumPy operation combining indexing and slicing operations onto the 3-channel array "vid_frame" to extract out a smaller 3-channel array called "roi_bgr", using "coordinates" (w1, h1) and (w2, h2) which define the top-left and bottom-right corners of a ROI centered in the video frame captured (see Lines 97-100 in Fig. 2.94 and the yellow box in the window labeled "Video Camera" shown in Fig. 2.91) (see https://docs.scipy.org/doc/numpy/reference/arrays.indexing.html for more details).
- Lines 216-217 uses the OpenCV method imshow() to display this "roi_bgr" in a small window located at position (50, 50) i.e. near the top left area of the computer display (see Fig. 2.91). Please note that imshow() uses a window wider that the ROI's width, thus creating the extra "blank-grey" area on the right of the displayed window. For more details regarding the use of imshow() and moveWindow() please visit this web link https://docs.opencv.org/4.0.1/d7/dfc/group__highgui.html#ga453d42fe4cb60e5723281a89973ee563)

```
214 # working on the ROI image/window
215     roi_bgr = vid_frame[h1:h2, w1:w2] # defining the ROI - numpy function
216     cv2.imshow('ROI_BGR', roi_bgr) # first ROI video window (BGR)
217     cv2.moveWindow('ROI_BGR', 50, 50)
218     roi_hsv = cv2.cvtColor(roi_bgr,cv2.COLOR_BGR2HSV)
219     cv2.imshow('ROI_HSV', roi_hsv) # second ROI video window (HSV)
220     cv2.moveWindow('ROI_HSV', 50, 200)
221     roi_mean, roi_std_dev = cv2.meanStdDev(roi_hsv)  # computing mean & std. dev.values of ROI (H,S,V)
222     print("ROI's mean (h,s,v) = " + str(int(roi_mean[0])) +
223           "\t" + str(int(roi_mean[1])) + "\t" + str(int(roi_mean[2])))
224     print("ROI's std dev (h,s,v) = " + str(int(roi_std_dev[0])) +
225           "\t" + str(int(roi_std_dev[1])) + "\t" + str(int(roi_std_dev[2])))
226     (h, s, v) = cv2.split(roi_hsv)    # split the ROI_HSV into its H-S-V image components
227     max_hsv[0] = np.max(h)
228     max_hsv[1] = np.max(s)
229     max_hsv[2] = np.max(v)
230     print("ROI's max (h,s,v) = " + str(max_hsv[0]) +
231           "\t" + str(max_hsv[1]) + "\t" + str(max_hsv[2]))
232     min_hsv[0] = np.min(h)
233     min_hsv[1] = np.min(s)
234     min_hsv[2] = np.min(v)
235     print("ROI's min (h,s,v) = " + str(min_hsv[0]) +
236           "\t" + str(min_hsv[1]) + "\t" + str(min_hsv[2]))
```

Fig. 2.100 Part 5 of Main Endless Loop in "AD_RemoteVision_HSV.py".

- Line 218 converts the ROI's BGR image into an HSV image named "roi_hsv" (see more details at https://docs.opencv.org/4.0.1/d8/d01/group__imgproc__color__conversions.html#ga397ae87e1288a81d2363b61574eb8cab and https://docs.opencv.org/master/df/d9d/tutorial_py_colorspaces.html).
- Lines 219-220 display "roi_hsv" in a window located at coordinates (50, 200) with respect to the computer display, i.e. right below the window for "roi_bgr" (see Fig. 2.91).
- Line 221 uses the OpenCV's method "meanStdDev()" to compute the respective means and standard deviations of all respective pixel values in each ROI image channel H-S-V, and save them in 2 1-D arrays named "roi_mean" and "roi_std_dev" which have 3 elements each (in the

HSV order of course). For more details please see web link at https://docs.opencv.org/4.0.1/d2/de8/group__core__array.html#ga846c858f4004d59493d7c6a4354b301d).

- Lines 222-225 print out the ROI's mean and standard deviation values found for each image channel Hue (element [0]), Saturation (element [1]) and Value (element [2]).
- Line 226 uses the OpenCV's method "split()" to unpack the 3-channel array "roi_hsv" into three single-channel arrays named "h", "s" and "v". See more details at web link (https://docs.opencv.org/master/d2/de8/group__core__array.html#ga8027f9deee1e42716be8039e5863fbd9).
- Lines 227-229 use the NumPy's method "max()" to find the maximum pixel values found in each respective single-channel array "h", "s" and "v" and save them in the respective element max_hsv[0], max_hsv[1] and max_hsv[2]. Line 230 prints out the current results found.
- Lines 232-236 does a similar job to find the respective minimum pixel values found in each (h, s, v) channel using the NumPy's method "min()".

Part 6 (see Fig. 1.101) wraps up operations for the Main Loop:

- Line 238 uses the OpenCV method rectangle() to draw a "yellow square ROI" inside the current "vid_frame", at locations defined by the two coordinates (w1, h1) and (w2, h2), using the color triplet (0, 255, 255) (remember OpenCV's BGR order here!), and the last argument "3" means a line thickness of 3. See link https://docs.opencv.org/master/d6/d6e/group__imgproc__draw.html#ga07d2f74cadcf8e305e810ce8eed13bc9 for more details about its usage.
- Line 239 then shows the current "vid_frame" with the yellow square embedded in its center area. Please note that the actual video frame was recorded (Line 213 of Fig. 2.99) **before** the "rectangle()" operation so the video file "video.avi" does not contain the yellow ROI.

```
237 # Adding the ROI rectangle in YELLOW onto the original video frame
238     cv2.rectangle(vid_frame, (w1, h1), (w2, h2), (0, 255, 255), 3)
239     cv2.imshow('Video Camera', vid_frame) # display modified captured frame
240
241 # End of MAIN LOOP
242
243 vid_cam.release()
244 vid_recorder.release()
245 cv2.destroyAllWindows()
```

Fig. 2.101 Part 6 of Main Endless Loop in "AD_RemoteVision_HSV.py".

If a break out of the Main Loop happens at run time, the program would flow to Lines 243-245 to terminate all OpenCV devices and windows, and the Python session terminates.

Video 2.23 shows the run time performance of "AD_RemoteVision_HSV.py" and illustrates how it can be used to record the various HSV data for Blue and Yellow objects of interest to be used in the next Python project "AD_CameraTracker.py".

Now we are ready to tackle the project "AD_CameraTracker.py" whereas the user can specify either "Blue" or "Yellow" objects for the robot Zeeb to autonomously search for and then, once found, approach it within a given distance controlled by the CM-150 built-in NIR sensors Left and Right (see Fig. 2.102).

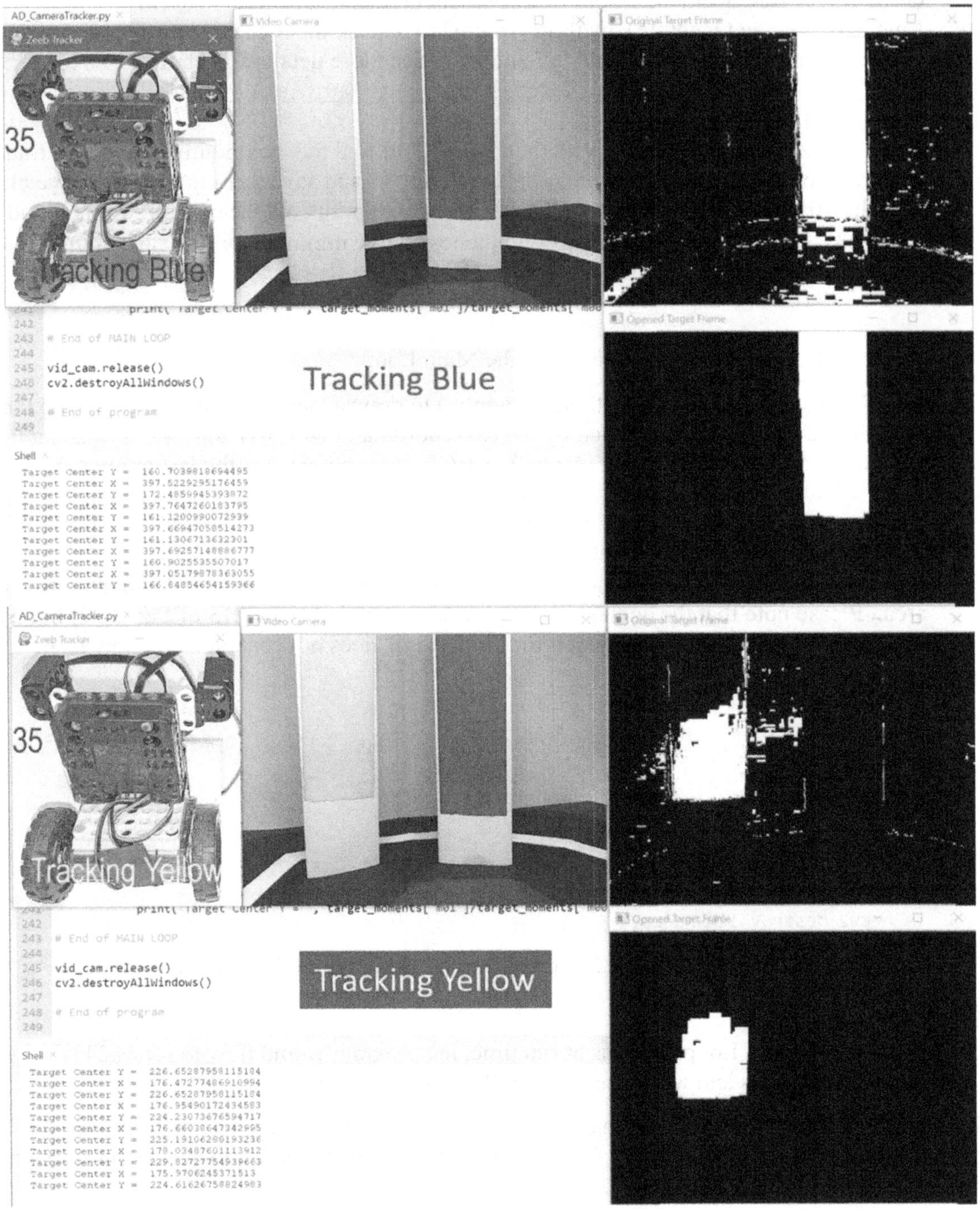

Fig. 2.102 "AD_CameraTracker.py" can be used to track "Blue" or "Yellow" objects.

Fig. 2.103 lists the beginning code for "AD_CameraTracker.py" where various Variables and Constants are initialized:

- Lines 67-74 are for the usual Variables and Messages.
- Lines 77-84 contain the Constants that were collected using "AD_RemoteVision_HSV.py" on the actual physical set up of the scene and objects to be tracked, it is recommended that the user keeps the same lighting setup between these two projects for best transfer data and runtime performances.

```
67 robot1 = "Zeeb"
68 robot_mode = 's' # Initial mode is 'stop'
69 speed = 35
70 r_speed = - speed
71 message1 = "1) Use Arrow keys to move robot around or 's' to stop the robot"
72 message2 = "2) Tap key 'i' to increase speed or 'd' to decrease speed"
73 message3 = "3) Tap key 'b' to track blue target or 'y' to track yellow target"
74 message4 = "4) AFTER target's color is chosen, tap key 'a' for autonomous search & approach"
75
76 # HSV Data previously collected from AD_RemoteVision_HSV
77 BLUE_OBJECT_MEAN = (111, 204, 232)
78 BLUE_OBJECT_STDDEV = (1, 2, 12)
79 BLUE_OBJECT_MIN = (109, 145, 145)
80 BLUE_OBJECT_MAX = (115, 220, 255)
81 YELLOW_OBJECT_MEAN = (22, 233, 241)
82 YELLOW_OBJECT_STDDEV = (1, 1, 7)
83 YELLOW_OBJECT_MIN = (21, 226, 227)
84 YELLOW_OBJECT_MAX = (23, 241, 255)
85
86 # Pixel values limits to be set for object to be tracked
87 obj_hsv_lo = [None] * 3
88 obj_hsv_hi = [None] * 3
89
90 # user needs to set up kernel size as needed for image processing - default is 20x20 pixels
91 kernel = np.ones((20,20),np.uint8)
```

Fig. 2.103 Initializations for Variables and Constants used in "AD_CameraTracker.py".

- Lines 87 and 88 define and initialize (to 0) two 1-D arrays for the Low and High HSV Limits which will be used for tracking the wanted object. Their actual values will be set by either Functions "set_blue_object()" or "set_yellow_object()" depending on the user's choice to be made at a later time (see Fig. 2.104).
- Line 91 defines a 2-D kernel or mask array of the size 20x20 filled with "1" for all elements. This kernel will be used in various image processing operations described in later code segments.

Fig. 2.104 lists the two Functions "set_blue_object()" or "set_yellow_object()" where the reader can see that low and high limits are set to the respective values "mean $\pm$ 2 * Std Dev".

```
def set_blue_object():
    global obj_hsv_lo, obj_hsv_hi
    for i in range(3):
        obj_hsv_lo[i] = BLUE_OBJECT_MEAN[i] - 2 * BLUE_OBJECT_STDDEV[i]
        obj_hsv_hi[i] = BLUE_OBJECT_MEAN[i] + 2 * BLUE_OBJECT_STDDEV[i]

def set_yellow_object():
    global obj_hsv_lo, obj_hsv_hi
    for i in range(3):
        obj_hsv_lo[i] = YELLOW_OBJECT_MEAN[i] - 2 * YELLOW_OBJECT_STDDEV[i]
        obj_hsv_hi[i] = YELLOW_OBJECT_MEAN[i] + 2 * YELLOW_OBJECT_STDDEV[i]
```

Fig. 2.104 Functions "set_blue_object()" and "set_yellow_object()" in "AD_CameraTracker.py".

Fig. 2.105 displays the usual preparation steps to get the camera ready for scene capture and video frame input check (Lines 122-131). Lines 134-135 define two new Variables needed for this project: "search_left_limit" and "search_right_limit" in the video frame coordinates system and please notice that only the width coordinate is of use to this project as the robot travels on a flat surface for this application. Line 121 initializes Variable "camera_mode" to NULL, it will be set with a proper value at a later time.

```
121 camera_mode = ''
122 vid_cam = cv2.VideoCapture(0)
123 if(not vid_cam.isOpened()):
124     print("Error connecting to camera - Exiting program")
125     vid_cam.release()
126     sys.exit()   # terminate program
127 
128 got_frame, vid_frame = vid_cam.read()  # picking up a new video frame
129 if not got_frame:   # error capturing video frame
130     print("No video frame captured - Exiting program")
131     sys.exit()  # terminate program
132 f_width = int(vid_cam.get(cv2.CAP_PROP_FRAME_WIDTH))
133 f_height = int(vid_cam.get(cv2.CAP_PROP_FRAME_HEIGHT))
134 search_left_limit = f_width/2 - 20
135 search_right_limit = f_width/2 + 20
136 print ("left_limit = " + str(search_left_limit))
137 print ("right_limit = " + str(search_right_limit))
```

Fig. 2.105 Preparation steps for camera usage in "AD_CameraTracker.py".

Fig. 2.106 defines the relevant text messages for this project which are "Tracking Blue", "Tracking Yellow" and "Autonomous S&A" (i.e. search and approach)

```
text_surface6 = font.render("Tracking Blue", True, (0, 0, 255))
x6 = ( SCREEN_SIZE[0] - text_surface6.get_width() ) / 2
y6 = 350
text_surface7 = font.render("Tracking Yellow", True, (255, 255, 0))
x7 = ( SCREEN_SIZE[0] - text_surface7.get_width() ) / 2
y7 = 350
text_surface8 = font.render("Autonomous S&A", True, (0, 255, 0))
x8 = ( SCREEN_SIZE[0] - text_surface8.get_width() ) / 2
y8 = ( SCREEN_SIZE[1] - text_surface8.get_height() ) / 2
```

Fig. 2.106 Relevant text messages in "AD_CameraTracker.py".

Fig. 2.107 shows Part 1 of the Main Algorithm designed around an Endless Loop as usual:

```
177 while True:
178 # Checking for user's keyboard inputs
179     for event in pygame.event.get():
180         if event.type == QUIT:  # when user closes Pygame window
181             stop(robot1)
182             vid_cam.release()
183             cv2.destroyAllWindows()
184             pygame.quit()
185             sys.exit()
```

Fig. 2.107 Part 1 of Main Algorithm in "AD_CameraTracker.py".

- Lines 180-185 check if the user closes the PyGame window (labeled "Zeeb Tracker" in Fig. 2.102) and act accordingly by stopping the robot (Line 181), by closing all OpenCV resources (Lines 182-183) and by quitting the PyGame module (Line 184) before a call to sys.exit() (Line 185).

Fig. 2.108 lists Part 2 which is a code segment used to check for a KEYDOWN event:

- If Key "b" is pushed, "camera_mode" is set to 'b' and Function "set_blue_object()" is called (Lines 187-189).
- If Key "y" is pushed, "camera_mode" is set to 'y' and Function "set_yellow_object()" is called (Lines 190-192).
- If Key "s" is pushed, the robot is stopped (Line 193-195). Similarly, the actions for Keys "i", "d" and arrow Keys had been explained earlier in previous projects (Lines 196-211).
- Lines 212-213 are activated when the user pushes on Key "a".

```
if event.type == KEYDOWN:
    if event.key == K_b:  # tracking blue
        camera_mode ='b'
        set_blue_object()
    if event.key == K_y:  # tracking yellow
        camera_mode ='y'
        set_yellow_object()
    if event.key == K_s:  # stop the robot
        robot_mode = 's'
        stop(robot1)
    if event.key == K_i:  # increase robot speed
        inc_speed(robot1)
    if event.key == K_d:  # decrease robot speed
        dec_speed(robot1)
    if event.key == K_UP:  # robot forward
        robot_mode = 'f'
        forward(robot1)
    if event.key == K_DOWN:  # robot backward
        robot_mode = 'b'
        backward(robot1)
    if event.key == K_LEFT:  # robot left
        robot_mode = 'l'
        left(robot1)
    if event.key == K_RIGHT:  # robot right
        robot_mode = 'r'
        right(robot1)
    if event.key == K_a:  # robot autonomous search & approach
        robot_mode = 'a'
```

Fig. 2.108 Part 2 of Main Algorithm in "AD_CameraTracker.py".

Part 3 of the Main Algorithm (Fig. 2.109) lists the various blitting tasks needed for this project based on the current value of Variable "robot_mode" which could be "f", "l", "r", "b", "s" or "a" (Lines 219-230). Similar tasks are also programmed for a possible value for Variable "camera_mode": "b" or "y" (Lines 232-235). Lastly, the current value for "speed" is displayed by Lines 237-238.

```
    screen = pygame.display.set_mode(SCREEN_SIZE, 0, 32)
    screen.blit(background, (0,0))

    if (robot_mode == 'f'):   # updating robot motion directions
        screen.blit(text_surface1, (x1, y1))
    elif (robot_mode == 'l'):
        screen.blit(text_surface2, (x2, y2))
    elif (robot_mode == 'r'):
        screen.blit(text_surface3, (x3, y3))
    elif (robot_mode == 'b'):
        screen.blit(text_surface4, (x4, y4))
    elif (robot_mode == 's'):
        screen.blit(text_surface5, (x5, y5))
    elif (robot_mode == 'a'):  # search & approach
        screen.blit(text_surface8, (x8, y8))

    if (camera_mode == 'b'):   # when tracking blue
        screen.blit(text_surface6, (x6, y6))
    elif (camera_mode == 'y'):  # when tracking yellow
        screen.blit(text_surface7, (x7, y7))

    text_surface9 = font.render(str(speed), True, (0, 0, 255))
    screen.blit(text_surface9, (0, 120))
    pygame.display.update()
```

Fig. 2.109 Part 3 of Main Algorithm in "AD_CameraTracker.py".

Part 4 (Fig. 2.110) shows the capture and display of a new video frame for each Main Loop iteration.

```
    got_frame, vid_frame = vid_cam.read()  # picking up a new video frame
    if not got_frame:   # error capturing video frame
        print("No video frame captured!")
        break   # get out of Main Loop
    cv2.imshow('Video Camera', vid_frame) # display original captured frame
```

Fig. 2.110 Part 4 of Main Algorithm in "AD_CameraTracker.py".

Part 5 (Fig. 2.111) lists the more interesting steps of processing the current video frame to look for the object of interest and to autonomously approach the found object:

- Part 5 is only activated when "camera_mode" is set to "b" or "y" (Line 248).

```
247 # Process image to track blue or yellow object
248     if (camera_mode == 'b' or camera_mode == 'y'):
249 # Convert vid_frame (BGR) to hsv_frame
250         hsv_frame = cv2.cvtColor(vid_frame,cv2.COLOR_BGR2HSV)
251 # working on the TARGET image/window
252         (h, s, v) = cv2.split(hsv_frame)     # split hsv_frame into its H-S-V image components
253         target_frame = cv2.inRange (h, obj_hsv_lo[0], obj_hsv_hi[0]) # using only Hue frame
254         cv2.imshow('Original Target Frame', target_frame) # display original target frame
255         target_opened = cv2.morphologyEx(target_frame, cv2.MORPH_OPEN, kernel)
256         cv2.imshow('Opened Target Frame', target_opened) # display opened target frame
257
258         target_moments = cv2.moments(target_opened)
259         if (target_moments['m00'] >= 50):
260             target_x = target_moments['m10']/target_moments['m00']
261             target_y = target_moments['m01']/target_moments['m00']
262             print('Target Center X = ', target_x)
263             print('Target Center Y = ', target_y)
264             if (robot_mode == 'a' and target_x > 10):
265                 if (target_x < search_left_limit):
266                     left(robot1)
267                 elif (target_x > search_right_limit):
268                     right(robot1)
269                 elif ((target_x >= search_left_limit) and (target_x <= search_right_limit)):
270                     speed = 35
271                     forward(robot1)
272                     sensors_data = ec.get_robot(robot1)["reporters"]
273                     ir_left = sensors_data["leftIR"]
274                     ir_right = sensors_data["rightIR"]
275                     if (ir_left > 30 or ir_right > 30):
276                         stop(robot1)
277                         break
```

Fig. 2.111 Part 5 of Main Algorithm in "AD_CameraTracker.py".

- The first step is to convert the original video frame from BGR to HSV color system (Line 250).
- The second step is to split the HSV video frame into its individual component "h", "s", and "v" and only "h" will be needed in this project (Line 252).
- The third step is to use the OpenCV method inRange() on the "h" frame (Line 253) to get a **new** "target_frame" which is a "binary" version of the "h" frame where:
    - H pixels that happen to have intensity values between the limits "obj_hsv_lo[0]" and "obj_hsv_hi[0]" are set to "255", i.e. they become "white".
    - For H pixels that fall outside of the range of those two limits they will be set to "0", i.e. they become "black".
    - See this web page for more details on Method "inRange()" (https://docs.opencv.org/master/d2/de8/group__core__array.html#ga48af0ab51e36436c5d04340e036ce981).
- Line 254 displays a typical "target_frame" obtained after the "inRange" step had been done for a Blue object (see Fig. 2.112). The reader can see that the main Blue object had been identified, but other smaller blobs were also identified by the "inRange" method. This was because we did not use the other limits on S and V with the "inRange" method (which would cost us more computer processing time). We were trading process speed for process accuracy because we

know that the robot would have to track quickly enough to provide real-time commands to its motors.

Fig. 2.112 Incoming "vid_frame" (left) and "target_frame" (right) obtained after an inRange() operation in "AD_CameraTracker.py".

- Fortunately, there is another speedy method that can be used on "target_frame" to get rid of most of the smaller blobs. It is an OpenCV morphological operator named "OPEN" (Line 255) (see link https://docs.opencv.org/master/d4/d86/group__imgproc__filter.html#gga7be549266bad7b2e6a04db49827f9f32a08d3cc3a2ace00cec488966d31fa29ea) which actually is a ERODE operation followed by a DILATE operation (https://docs.opencv.org/master/d4/d86/group__imgproc__filter.html#gaeb1e0c1033e3f6b891a25d0511362aeb). Its result is shown in Fig. 2.113 (Line 256) and this result depends on the size of the kernel used (Line 91 of Fig. 2.103), thus readers may have to adjust this parameter for their own situations. As we only need to know the width coordinate of the CG of the white blob in Fig. 2.113 in order to steer the robot left or right to align it with the Blue object in the physical realm, Fig. 2.113 tells us that we are on the right "track"!

Fig. 2.113 Result of an "OPEN" operation on the "target_frame" shown in Fig. 2.112.

- The width coordinate of the CG of the white blob in Fig. 2.113 (i.e. Variable "target_x") can be determined with the use of the OpenCV Class called "moments()" (https://docs.opencv.org/master/d8/d23/classcv_1_1Moments.html) on the binary frame "target_opened" (Line 258). The "moments()" class returns many parameters stored in the dictionary named "target_moments", and we will only need to use two type of parameters from it:
    - The "m00" item represents the total number of white pixels occupied by the white blob in the "target_opened" frame (or so called zeroth order moment). If "m00" is zero or smaller than 50, we shall assume that no valid target had been found (Line 259) and this control algorithm is by-passed back out to the Main Endless Loop.
    - The "m10" item represents the first moment of those white pixels with respect to the width coordinate wherever a given white pixel is found within the "target_opened" frame.
    - When (m00 >= 50), the expression (m10 / m00) can be computed and would give us the width coordinate of the CG (or geometrical barycenter) of the white blob (i.e. "target_x" in Line 260), wherever it may appear as the camera pans across the physical scene. This leads to a very simple control algorithm to align the robot/camera with the wanted target described in a later paragraph.
    - In Fig. 2.111, the height-based barycenter coordinate "target_y" is also computed and printed, but it is just for informational use only.

- Lines 264-277 describe the target alignment and approach algorithm for the robot:
    - The IF condition used in Line 264 implies that a little bit of the user-designated object needs to be within the camera viewport at the start of this algorithm, i.e. target_x > 10 (*the reader could improve on this algorithm by making the robot spinning left or right when no target is within the camera's viewport – a similar approach was used for the Zeeb-Zob Dowel Game project*). When this condition is True, "target_x" is further checked to see if "target_x" is less than "search_left_limit" (Line 267). If this latter condition is True, then the target is on the physical left of the robot, so it should spin left

to bring more of the target into the camera's viewpoint (Line 266). However, if Line 265 turns out to be False, then Line 267 needs to be checked next, to see if (target_x > search_right_limit). If this condition is True, then the target is on the physical right of the robot, then it should spin right to get more of the target into the camera viewpoint (Line 268).

- The two robot actions described above should "wiggle" the robot right and left as needed until the condition in Line 269 becomes True, i.e. the robot is acceptably aligned with the target (i.e. target's CG is within $\pm$ 20 pixels on either side of the vertical middle line across the video frame), then the robot is set to go forward at speed 35 (Lines 270-271). Next, current data inputs from the built-in NIR Left and Right Sensors are monitored (Lines 272-275). When either "ir_left" or "ir_right" is greater than "30" (Line 275), the physical distance between the physical target and the robot is considered close enough, and the algorithm would stop the robot (Line 276) and break out of the Main Endless Loop (Line 277). Next, the Python session would terminate as normal.

Video 2.24 showcases a typical run-time performance of this project "AD_CameraTracker.py" which is quite satisfactory, the author believes.

As a conclusion to this book, the author hopes to have shown that the DREAM system is quite capable to help a user build a very solid knowledge base for integrating sensor characteristics with actuator capabilities and underlying communications constraints to design and control a single or multiple robotics system with joy and confidence!

2.7 References

Bouras AS, Ainarozidou LV (2017) Python for Tweens and Teens: Learn Computational And Algorithmic Thinking. Published by Amazon Digital Services LLC.

Escrivá DM and Mendonça VG (2018) Learn OpenCV 4 by Building Projects. Packt, Birmingham, UK.

Ford JL Junior (2014) Scratch 2.0 programming for teens. CENGAGE Learning PTR, Boston USA.

Hellmann D (2017) The Python 3 Standard Library by Example. Addison-Wesley Professional, Boston USA.

Kaehler A and Bradski G (2017) Learning OpenCV 3: Computer Vision in C++ with the OpenCV Library, 1st Edition. O'Reilly Media, Sebastopol USA.

Kinsley H and McGugan W (2015) Beginning Python Games Development with PyGame, 2nd Edition. Apress, New York.

Matthes E (2019) Python Crash Course, 2nd Edition. No Starch Press, San Francisco USA.

Nguyen Q (2018) Mastering Concurrency in Python. Packt, Birmingham, UK.

Spizhevoy A and Rybnikov A (2018) OpenCV 3 Computer Vision with Python Cookbook: Leverage the power of OpenCV 3 and Python to build computer vision applications. Packt, Birmingham, UK.

Summerfield M (2010) Programming in Python 3, 2nd Edition. Addison-Wesley Professional, Boston USA.

Thai CN (2017) Exploring Robotics with ROBOTIS Systems (2nd Edition). Springer-Nature Publishers, Cham Switzerland.

Thai CN (2018) Learning Robotics with ROBOTIS DREAM System. CNT Robotics LLC Publisher. Printed by Amazon Create Space.

Vlieg EA (2016) Scratch by Example: Programming for All Ages. Apress Media, LLC, New York USA

Warner TL (2015) Teach yourself Scratch 2.0 in 24 hours. SAMS, Indianapolis USA.

Made in the USA
Monee, IL
05 November 2026

67386704R00096